Fundamentals of Industrial Instrumentation and Process Control

William C. Dunn

McGraw-Hill

New York   Chicago   San Francisco   Lisbon   London   Madrid
Mexico City   Milan   New Delhi   San Juan   Seoul
Singapore   Sydney   Toronto

Library of Congress Cataloging-in-Publication Data

Dunn, William (William Charles), date.
  Fundamentals of industrial instrumentation and process control / William Dunn.
    p.   cm.
  ISBN 0-07-145735-6
  1. Engineering instruments.   I. Title.

TA165.D86   2005
670.42—dc22

2005041571

Copyright © 2005 by The McGraw-Hill Companies, Inc. All rights reserved. Printed in the United States of America. Except as permitted under the United States Copyright Act of 1976, no part of this publication may be reproduced or distributed in any form or by any means, or stored in a data base or retrieval system, without the prior written permission of the publisher.

1 2 3 4 5 6 7 8 9 0   DOC/DOC   0 1 0 9 8 7 6 5

ISBN 0-07-145735-6

The sponsoring editor for this book was Kenneth P. McCombs and the production supervisor was Sherri Souffrance. It was set in Century Schoolbook by International Typesetting and Composition. The art director for the cover was Anthony Landi.

Printed and bound by RR Donnelley.

 This book was printed on recycled, acid-free paper containing a minimum of 50% recycled, de-inked fiber.

McGraw-Hill books are available at special quantity discounts to use as premiums and sales promotions, or for use in corporate training programs. For more information, please write to the Director of Special Sales, McGraw-Hill Professional, Two Penn Plaza, New York, NY 10121-2298. Or contact your local bookstore.

Information contained in this work has been obtained by The McGraw-Hill Companies, Inc. ("McGraw-Hill") from sources believed to be reliable. However, neither McGraw-Hill nor its authors guarantee the accuracy or completeness of any information published herein, and neither McGraw-Hill nor its authors shall be responsible for any errors, omissions, or damages arising out of use of this information. This work is published with the understanding that McGraw-Hill and its authors are supplying information but are not attempting to render engineering or other professional services. If such services are required, the assistance of an appropriate professional should be sought.

To my wife Nadine for her patience, understanding, and many helpful suggestions during the writing of this text

Contents

Preface   xiii

Chapter 1. Introduction and Review   1

    Chapter Objectives   1
1.1 Introduction   1
1.2 Process Control   2
1.3 Definitions of the Elements in a Control Loop   3
1.4 Process Facility Considerations   6
1.5 Units and Standards   7
1.6 Instrument Parameters   9
    Summary   13
    Problems   13

Chapter 2. Basic Electrical Components   15

    Chapter Objectives   15
2.1 Introduction   15
2.2 Resistance   16
    2.2.1 Resistor formulas   17
    2.2.2 Resistor combinations   19
    2.2.3 Resistive sensors   23
2.3 Capacitance   24
    2.3.1 Capacitor formulas   24
    2.3.2 Capacitor combinations   25
2.4 Inductance   26
    2.4.1 Inductor formulas   26
    2.4.2 Inductor combinations   27
    Summary   27
    Problems   28

Chapter 3. AC Electricity   31

    Chapter Objectives   31
3.1 Introduction   31
3.2 Circuits with R, L, and C 32

	3.2.1	Voltage step	32
	3.2.2	Time constants	33
	3.2.3	Phase change	35
3.3	*RC* Filters		38
3.4	AC Bridges		39
3.5	Magnetic Forces		40
	3.5.1	Magnetic fields	40
	3.5.2	Analog meter	42
	3.5.3	Electromechanical devices	43
	Summary		44
	Problems		45

Chapter 4. Electronics — 47

	Chapter Objectives		47
4.1	Introduction		48
4.2	Analog Circuits		48
	4.2.1	Discrete amplifiers	48
	4.2.2	Operational amplifiers	49
	4.2.3	Current amplifiers	53
	4.2.4	Differential amplifiers	54
	4.2.5	Buffer amplifiers	55
	4.2.6	Nonlinear amplifiers	56
	4.2.7	Instrument amplifier	56
	4.2.8	Amplifier applications	57
4.3	Digital Circuits		58
	4.3.1	Digital signals	58
	4.3.2	Binary numbers	58
	4.3.3	Logic circuits	60
	4.3.4	Analog-to-digital conversion	61
4.4	Circuit Considerations		63
	Summary		63
	Problems		64

Chapter 5. Pressure — 67

	Chapter Objectives		67
5.1	Introduction		67
5.2	Basic Terms		68
5.3	Pressure Measurement		69
5.4	Pressure Formulas		70
5.5	Measuring Instruments		73
	5.5.1	Manometers	73
	5.5.2	Diaphragms, capsules, and bellows	75
	5.5.3	Bourdon tubes	77
	5.5.4	Other pressure sensors	79
	5.5.5	Vacuum instruments	79
5.6	Application Considerations		80
	5.6.1	Selection	80
	5.6.2	Installation	80
	5.6.3	Calibration	81
	Summary		81
	Problems		82

Chapter 6. Level — 85

    Chapter Objectives — 85
6.1 Introduction — 85
6.2 Level Formulas — 86
6.3 Level Sensing Devices — 87
    6.3.1 Direct level sensing — 88
    6.3.2 Indirect level sensing — 92
6.4 Application Considerations — 95
    Summary — 97
    Problems — 97

Chapter 7. Flow — 99

    Chapter Objectives — 99
7.1 Introduction — 99
7.2 Basic Terms — 100
7.3 Flow Formulas — 102
    7.3.1 Continuity equation — 102
    7.3.2 Bernoulli equation — 103
    7.3.3 Flow losses — 105
7.4 Flow Measurement Instruments — 107
    7.4.1 Flow rate — 107
    7.4.2 Total flow — 111
    7.4.3 Mass flow — 112
    7.4.4 Dry particulate flow rate — 113
    7.4.5 Open channel flow — 113
7.5 Application Considerations — 114
    7.5.1 Selection — 114
    7.5.2 Installation — 115
    7.5.3 Calibration — 115
    Summary — 115
    Problems — 116

Chapter 8. Temperature and Heat — 119

    Chapter Objectives — 119
8.1 Introduction — 119
8.2 Basic Terms — 120
    8.2.1 Temperature definitions — 120
    8.2.2 Heat definitions — 121
    8.2.3 Thermal expansion definitions — 123
8.3 Temperature and Heat Formulas — 124
    8.3.1 Temperature — 124
    8.3.2 Heat transfer — 124
    8.3.3 Thermal expansion — 126
8.4 Temperature Measuring Devices — 127
    8.4.1 Thermometers — 127
    8.4.2 Pressure-spring thermometers — 129
    8.4.3 Resistance temperature devices — 130
    8.4.4 Thermistors — 131
    8.4.5 Thermocouples — 131
    8.4.6 Semiconductors — 133

8.5	Application Considerations	134
	8.5.1	Selection	134
	8.5.2	Range and accuracy	134
	8.5.3	Thermal time constant	134
	8.5.4	Installation	137
	8.5.5	Calibration	137
	8.5.6	Protection	137
	Summary		138
	Problems		138

Chapter 9. Humidity, Density, Viscosity, and pH — 141

	Chapter Objectives	141
9.1	Introduction	141
9.2	Humidity	142
	9.2.1	Humidity definitions	142
	9.2.2	Humidity measuring devices	146
9.3	Density and Specific Gravity	149
	9.3.1	Basic terms	149
	9.3.2	Density measuring devices	150
	9.3.3	Density application considerations	153
9.4	Viscosity	153
	9.4.1	Basic terms	153
	9.4.2	Viscosity measuring instruments	154
9.5	pH Measurements	155
	9.5.1	Basic terms	155
	9.5.2	pH measuring devices	156
	9.5.3	pH application considerations	156
	Summary		157
	Problems		158

Chapter 10. Other Sensors — 161

	Chapter Objectives	161
10.1	Introduction	161
10.2	Position and Motion Sensing	161
	10.2.1	Basic position definitions	161
	10.2.2	Position and motion measuring devices	163
	10.2.3	Position application consideration	166
10.3	Force, Torque, and Load Cells	166
	10.3.1	Basic definitions of force and torque	166
	10.3.2	Force and torque measuring devices	167
	10.3.3	Force and torque application considerations	170
10.4	Smoke and Chemical Sensors	170
	10.4.1	Smoke and chemical measuring devices	171
	10.4.2	Smoke and chemical application consideration	171
10.5	Sound and Light	171
	10.5.1	Sound and light formulas	171
	10.5.2	Sound and light measuring devices	173
	10.5.3	Light sources	174
	10.5.4	Sound and light application considerations	174
	Summary		176
	Problems		176

Chapter 11. Actuators and Control — 179

- Chapter Objectives — 179
- 11.1 Introduction — 179
- 11.2 Pressure Controllers — 180
  - 11.2.1 Regulators — 180
  - 11.2.2 Safety valves — 182
  - 11.2.3 Level regulators — 182
- 11.3 Flow Control Actuators — 183
  - 11.3.1 Globe valve — 183
  - 11.3.2 Butterfly valve — 185
  - 11.3.3 Other valve types — 185
  - 11.3.4 Valve characteristics — 186
  - 11.3.5 Valve fail safe — 187
- 11.4 Power Control — 188
  - 11.4.1 Electronic devices — 188
  - 11.4.2 Magnetic control devices — 193
- 11.5 Motors — 195
  - 11.5.1 Servo motors — 195
  - 11.5.2 Stepper motors — 195
  - 11.5.3 Valve position feedback — 196
  - 11.5.4 Pneumatic feedback — 196
- 11.6 Application Considerations — 196
  - 11.6.1 Valves — 196
  - 11.6.2 Power devices — 197
- Summary — 198
- Problems — 198

Chapter 12. Signal Conditioning — 201

- Chapter Objectives — 201
- 12.1 Introduction — 201
- 12.2 Conditioning — 202
  - 12.2.1 Characteristics — 202
  - 12.2.2 Linearization — 204
  - 12.2.3 Temperature correction — 205
- 12.3 Pneumatic Signal Conditioning — 205
- 12.4 Visual Display Conditioning — 206
  - 12.4.1 Direct reading sensors — 206
- 12.5 Electrical Signal Conditioning — 207
  - 12.5.1 Linear sensors — 208
  - 12.5.2 Float sensors — 208
  - 12.5.3 Strain gauge sensors — 211
  - 12.5.4 Capacitive sensors — 212
  - 12.5.5 Resistance sensors — 213
  - 12.5.6 Magnetic sensors — 214
  - 12.5.7 Thermocouple sensors — 215
  - 12.5.8 Other sensors — 215
- 12.6 A-D Conversion — 216
- Summary — 216
- Problems — 216

Chapter 13. Signal Transmission — 219

    Chapter Objectives — 219
- 13.1 Introduction — 220
- 13.2 Pneumatic Transmission — 220
- 13.3 Analog Transmission — 220
  - 13.3.1 Noise considerations — 220
  - 13.3.2 Voltage signals — 222
  - 13.3.3 Current signals — 223
  - 13.3.4 Signal conversion — 223
  - 13.3.5 Thermocouples — 224
  - 13.3.6 Resistance temperature devices — 225
- 13.4 Digital Transmission — 226
  - 13.4.1 Transmission standards — 226
  - 13.4.2 Smart sensors — 227
  - 13.4.3 Foundation Fieldbus and Profibus — 229
- 13.5 Controller — 230
  - 13.5.1 Controller operation — 231
  - 13.5.2 Ladder diagrams — 232
- 13.6 Digital-to-Analog Conversion — 235
  - 13.6.1 Digital-to-analog converters — 235
  - 13.6.2 Pulse width modulation — 236
- 13.7 Telemetry — 237
  - 13.7.1 Width modulation — 237
  - 13.7.2 Frequency modulation — 238
- Summary — 239
- Problems — 239

Chapter 14. Process Control — 241

    Chapter Objectives — 241
- 14.1 Introduction — 241
- 14.2 Basic Terms — 242
- 14.3 Control Modes — 243
  - 14.3.1 ON/OFF action — 243
  - 14.3.2 Differential action — 244
  - 14.3.3 Proportional action — 244
  - 14.3.4 Derivative action — 246
  - 14.3.5 Integral action — 247
  - 14.3.6 PID action — 248
- 14.4 Implementation of Control Loops — 249
  - 14.4.1 ON/OFF action pneumatic controller — 249
  - 14.4.2 ON/OFF action electrical controller — 250
  - 14.4.3 PID action pneumatic controller — 251
  - 14.4.4 PID action control circuits — 252
  - 14.4.5 PID electronic controller — 254
- 14.5 Digital Controllers — 256
- Summary — 257
- Problems — 257

Chapter 15. Documentation and Symbols — 259

    Chapter Objectives — 259
- 15.1 Introduction — 259

15.2	System Documentation	260
	15.2.1 Alarm and trip systems	260
	15.2.2 Alarm and trip documentation	261
	15.2.3 PLC documentation	261
15.3	Pipe and Identification Diagrams	262
	15.3.1 Standardization	262
	15.3.2 Interconnections	262
	15.3.3 Instrument symbols	263
	15.3.4 Instrument identification	264
15.4	Functional Symbols	266
	15.4.1 Actuators	266
	15.4.2 Primary elements	266
	15.4.3 Regulators	267
	15.4.4 Math functions	267
15.5	P and ID Drawings	267
	Summary	269
	Problems	271

Appendix A. Units   273

Appendix B. Thermocouple Tables   277

Appendix C. References and Information Resources   279

Appendix D. Abbreviations   283

Glossary   287
Answers to Odd-Numbered Questions   297
Index   311

ABOUT THE AUTHOR

William Dunn has B.Sc. in physics from the University of London, graduating with honors, he also has a B.S.E.E. He has over 40 years industrial experience in management, marketing support, customer interfacing, and advanced product development in systems and microelectronic and micromachined sensor development. Most recently he taught industrial instrumentation, and digital logic at Ouachita Technical College as an adjunct professor. Previously he was with Motorola Semiconductor Product Sector working in advanced product development, designing micromachined sensors and transducers. He holds some 15 patents in sensor design, and has presented some 20 technical papers in sensor design and application.

Preface

Instrumentation and process control can be traced back many millennia. Some of the early examples are the process of making fire and instruments using the sun and stars, such as Stonehenge. The evolution of instrumentation and process control has undergone several industrial revolutions leading to the complexities of modern day microprocessor-controlled processing. Today's technological evolution has made it possible to measure parameters deemed impossible only a few years ago. Improvements in accuracy, tighter control, and waste reduction have also been achieved.

This book was specifically written as an introduction to modern day industrial instrumentation and process control for the two-year technical, vocational, or degree student, and as a reference manual for managers, engineers, and technicians working in the field of instrumentation and process control. It is anticipated that the prospective student will have a basic understanding of mathematics, electricity, and physics. This course should adequately prepare a prospective technician, or serve as an introduction for a prospective engineer wishing to get a solid basic understanding of instrumentation and process control.

Instrumentation and process control involve a wide range of technologies and sciences, and they are used in an unprecedented number of applications. Examples range from the control of heating, cooling, and hot water systems in homes and offices to chemical and automotive instrumentation and process control. This book is designed to cover all aspects of industrial instrumentation, such as sensing a wide range of variables, the transmission and recording of the sensed signal, controllers for signal evaluation, and the control of the manufacturing process for a quality and uniform product.

Chapter 1 gives an introduction to industrial instrumentation. Chapters 2 through 4 refresh the student's knowledge of basic electricity and introduce electrical circuits for use in instrumentation. Sensors and their use in the measurement of a wide variety of physical variables—such as level, pressure, flow, temperature, humidity, and mechanical measurements—are discussed in Chapters 5 through 10. The use of regulators and actuators for controlling pressure, flow, and the control of the input variables to a process are discussed in

Chapter 11. Electronics is the medium for sensor signal amplification, conditioning, transmission, and control. These functions are presented as they apply to process control in Chapters 12 through 14. Finally, in Chapter 15, documentation as applied to instrumentation and control is introduced, together with standard symbols recommended by the Instrument Society of America (ISA) for use in instrumentation control diagrams.

The primary reason for writing this book was that the author felt that there was no clear, concise, and up-to-date book for prospective technicians and engineers which could help them understand the basics of instrumentation and process control. Every effort has been made to ensure that the book is accurate, easily readable, and understandable.

Both engineering and scientific units are discussed in the book. Each chapter contains worked examples for clarification, with exercise problems at the end of each chapter. A glossary and answers to the odd-numbered questions are given at the end of the book.

William C. Dunn

Chapter 1

Introduction and Review

Chapter Objectives

This chapter will introduce you to instrumentation, the various measurement units used, and the reason why process control relies extensively on instrumentation. It will help you become familiar with instrument terminology and standards.

This chapter discusses

- The basics of a process control loop
- The elements in a control loop
- The difference between the various types of variables
- Considerations in a process facility
- Units, standards, and prefixes used in parameter measurements
- Comparison of the English and the SI units of measurement
- Instrument accuracy and parameters that affect an instrument's performance

1.1 Introduction

Instrumentation is the basis for process control in industry. However, it comes in many forms from domestic water heaters and HVAC, where the variable temperature is measured and used to control gas, oil, or electricity flow to the water heater, or heating system, or electricity to the compressor for refrigeration, to complex industrial process control applications such as used in the petroleum or chemical industry.

In industrial control a wide number of variables, from temperature, flow, and pressure to time and distance, can be sensed simultaneously. All of these can be interdependent variables in a single process requiring complex microprocessor systems for total control. Due to the rapid advances in technology, instruments in use today may be obsolete tomorrow, as new and more efficient measurement techniques are constantly being introduced. These changes are being driven by the need for higher accuracy, quality, precision, and performance. To measure parameters accurately, techniques have been developed that were thought impossible only a few years ago.

1.2 Process Control

In order to produce a product with consistently high quality, tight process control is necessary. A simple-to-understand example of process control would be the supply of water to a number of cleaning stations, where the water temperature needs to be kept constant in spite of the demand. A simple control block is shown in Fig. 1.1a, steam and cold water are fed into a heat exchanger, where heat from the steam is used to bring the cold water to the required working temperature. A thermometer is used to measure the temperature of the water (the measured variable) from the process or exchanger. The temperature is observed by an operator who adjusts the flow of steam (the manipulated variable) into the heat exchanger to keep the water flowing from the heat exchanger at the constant set temperature. This operation is referred to as process control, and in practice would be automated as shown in Fig. 1.1b.

Process control is the automatic control of an output variable by sensing the amplitude of the output parameter from the process and comparing it to the desired or set level and feeding an error signal back to control an input variable—in this case steam. See Fig. 1.1b. A temperature sensor attached to the outlet pipe senses the temperature of the water flowing. As the demand for hot water increases or decreases, a change in the water temperature is sensed and converted to an electrical signal, amplified, and sent to a controller that evaluates the signal and sends a correction signal to an actuator. The actuator adjusts the flow of steam to the heat exchanger to keep the temperature of the water at its predetermined value.

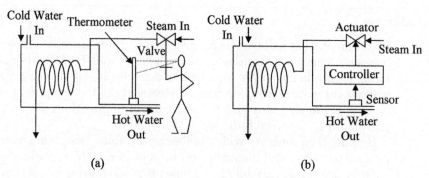

Figure 1.1 Process control (*a*) shows the manual control of a simple heat exchanger process loop and (*b*) automatic control of a heat exchanger process loop.

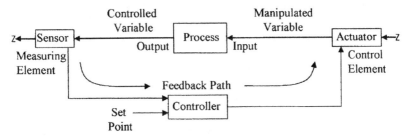

Figure 1.2 Block diagram of a process control loop.

The diagram in Fig. 1.1b is an oversimplified feedback loop and is expanded in Fig. 1.2. In any process there are a number of inputs, i.e., from chemicals to solid goods. These are manipulated in the process and a new chemical or component emerges at the output. The controlled inputs to the process and the measured output parameters from the process are called variables.

In a process-control facility the controller is not necessarily limited to one variable, but can measure and control many variables. A good example of the measurement and control of multivariables that we encounter on a daily basis is given by the processor in the automobile engine. Figure 1.3 lists some of the functions performed by the engine processor. Most of the controlled variables are six or eight devices depending on the number of cylinders in the engine. The engine processor has to perform all these functions in approximately 5 ms. This example of engine control can be related to the operations carried out in a process-control operation.

1.3 Definitions of the Elements in a Control Loop

Figure 1.4 breaks down the individual elements of the blocks in a process-control loop. The measuring element consists of a sensor, a transducer, and a transmitter with its own regulated power supply. The control element has an actuator, a power control circuit, and its own power supply. The controller has a processor with a

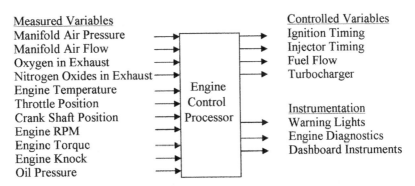

Figure 1.3 Automotive engine showing some of the measured and controlled variables.

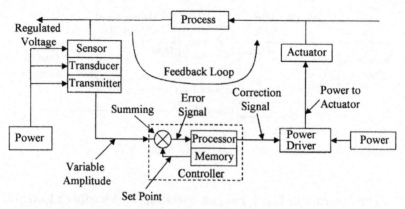

Figure 1.4 Block diagram of the elements that make up the feedback path in a process-control loop.

memory and a summing circuit to compare the set point to the sensed signal so that it can generate an error signal. The processor then uses the error signal to generate a correction signal to control the actuator and the input variable. The function and operation of the blocks in different types of applications will be discussed in Chaps. 11, 12, and 14. The definition of these blocks is given as follows:

Feedback loop is the signal path from the output back to the input to correct for any variation between the output level from the set level. In other words, the output of a process is being continually monitored, the error between the set point and the output parameter is determined, and a correction signal is then sent back to one of the process inputs to correct for changes in the measured output parameter.

Controlled or measured variable is the monitored output variable from a process. The value of the monitored output parameter is normally held within tight given limits.

Manipulated variable is the input variable or parameter to a process that is varied by a control signal from the processor to an actuator. By changing the input variable the value of the measured variable can be controlled.

Set point is the desired value of the output parameter or variable being monitored by a sensor. Any deviation from this value will generate an error signal.

Instrument is the name of any of the various device types for indicating or measuring physical quantities or conditions, performance, position, direction, and the like.

Sensors are devices that can detect physical variables, such as temperature, light intensity, or motion, and have the ability to give a measurable output that varies in relation to the amplitude of the physical variable. The human body has sensors in the fingers that can detect surface roughness, temperature, and force. A thermometer is a good example of a line-of-sight sensor, in that it will give an accurate visual indication of temperature. In other sensors such as a diaphragm pressure sensor, a strain transducer may be required to convert the deformation of the diaphragm into an electrical or pneumatic signal before it can be measured.

Transducers are devices that can change one form of energy to another, e.g., a resistance thermometer converts temperature into electrical resistance, or a thermocouple converts temperature into voltage. Both of these devices give an output that is proportional to the temperature. Many transducers are grouped under the heading of sensors.

Converters are devices that are used to change the format of a signal without changing the energy form, i.e., a change from a voltage to a current signal.

Actuators are devices that are used to control an input variable in response to a signal from a controller. A typical actuator will be a flow-control valve that can control the rate of flow of a fluid in proportion to the amplitude of an electrical signal from the controller. Other types of actuators are magnetic relays that turn electrical power on and off. Examples are actuators that control power to the fans and compressor in an air-conditioning system in response to signals from the room temperature sensors.

Controllers are devices that monitor signals from transducers and take the necessary action to keep the process within specified limits according to a predefined program by activating and controlling the necessary actuators.

Programmable logic controllers (PLC) are used in process-control applications, and are microprocessor-based systems. Small systems have the ability to monitor several variables and control several actuators, with the capability of being expanded to monitor 60 or 70 variables and control a corresponding number of actuators, as may be required in a petrochemical refinery. PLCs, which have the ability to use analog or digital input information and output analog or digital control signals, can communicate globally with other controllers, are easily programmed on line or off line, and supply an unprecedented amount of data and information to the operator. Ladder networks are normally used to program the controllers.

An error signal is the difference between the set point and the amplitude of the measured variable.

A correction signal is the signal used to control power to the actuator to set the level of the input variable.

Transmitters are devices used to amplify and format signals so that they are suitable for transmission over long distances with zero or minimal loss of information. The transmitted signal can be in one of the several formats, i.e., pneumatic, digital, analog voltage, analog current, or as a radio frequency (RF) modulated signal. Digital transmission is preferred in newer systems because the controller is a digital system, and as analog signals can be accurately digitized, digital signals can be transmitted without loss of information. The controller compares the amplitude of the signal from the sensor to a predetermined set

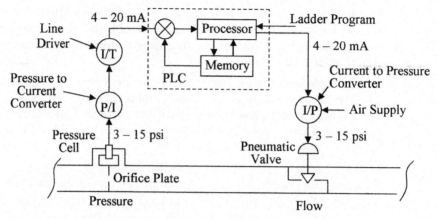

Figure 1.5 Process control with a flow regulator for use in Example 1.1.

point, which in Fig. 1.1*b* is the amplitude of the signal of the hot water sensor. The controller will then send a signal that is proportional to the difference between the reference and the transmitted signal to the actuator telling the actuator to open or close the valve controlling the flow of steam to adjust the temperature of the water to its set value.

Example 1.1 Figure 1.5 shows the block diagram of a closed-loop flow control system. Identify the following elements: (*a*) the sensor, (*b*) the transducer, (*c*) the actuator, (*d*) the transmitter, (*e*) the controller, (*f*) the manipulated variable, and (*g*) the measured variable.

(*a*) The sensor is labeled pressure cell in the diagram. (*b*) The transducer is labeled converter. There are two transducers—one for converting pressure to current and the other for converting current to pressure to operate the actuator. (*c*) The actuator in this case is the pneumatic valve. (*d*) The transmitter is the line driver. (*e*) The controller is labeled PLC. (*f*) The manipulated variable is the differential pressure developed by the fluid flowing through the orifice plate constriction. (*g*) The controlled variable is the flow rate of the liquid.

Simple and ideal process-control systems have been discussed. In practical process control the scenarios are much more complex with many scenarios and variables, such as stability, reaction time, and accuracy to be considered. Many of the basic problems are discussed in the following chapters.

1.4 Process Facility Considerations

The process facility has a number of basic requirements including safety precautions and well-regulated, reliable electrical, water, and air supplies.

An electrical supply is required for all control systems and must meet all standards in force at the plant. The integrity of the electrical supply is most important. Many facilities have backup systems to provide an uninterruptible power supply (UPS) to take over in case of loss of external power. Power failure can mean plant shutdown and the loss of complete production runs. An isolating transformer should be used in the power supply lines to prevent electromagnetic interference

(EMI) generated by motors, contactors, relays, and so on from traveling through the power lines and affecting sensitive electronic control instruments.

Grounding is a very important consideration in a facility for safety reasons. Any variations in the ground potential between electronic equipment can cause large errors in signal levels. Each piece of equipment should be connected to a heavy copper bus that is properly grounded. Ground loops should also be avoided by grounding cable screens and signal return lines at one end only. In some cases it may be necessary to use signal isolators to alleviate grounding problems in electronic devices and equipment.

An air supply is required to drive pneumatic actuators in most facilities. Instrument air in pneumatic equipment must meet quality standards, the air must be dirt, oil, contaminant, and moisture free. Frozen moisture, dirt, and the like can fully or partially block narrowed sections and nozzles, giving false readings or complete equipment failure. Air compressors are fitted with air dryers and filters, and have a reservoir tank with a capacity large enough for several minutes' supply in case of system failure. Dry, clean air is supplied at a pressure of 90 psig (630 kPa·g) and with a dew point of 20°F (10°C) below the minimum winter operating temperature at atmospheric pressure. Additional information on the quality of instrument air can be found in ANSI/ISA-7.0.01-1996, *Quality Standard for Instrument Air*.

Water supply is required in many cleaning and cooling operations, and for steam generation. Domestic water supplies contain large quantities of particulates and impurities, and may be satisfactory for cooling, but are not suitable for most cleaning operations. Filtering and other similar processes can remove some of the contaminants making the water suitable for some cleaning operations, but for ultrapure water a reverse osmosis system may be required.

Installation and maintenance must be considered when locating instruments, valves and so on. Each device must be easily accessible for maintenance and inspection. It may also be necessary to install hand-operated valves so that equipment can be replaced or serviced without complete plant shutdown. It may be necessary to contract out maintenance of certain equipment or have the vendor install equipment, if the necessary skills are not available in-house.

Safety is a top priority in a facility. The correct material must be used in container construction, plumbing, seals, and gaskets to prevent corrosion and failure leading to leakage and spills of hazardous materials. All electrical equipment must be properly installed to code with breakers. Electrical systems must have the correct fire retardant for use in case of electrical fires. More information can be found in ANSI/ISA-12.01.01-1999, *Definitions and Information Pertaining to Electrical Instruments in Hazardous Locations*.

1.5 Units and Standards

As with all disciplines, a set of standards has evolved over the years to ensure consistency and avoid confusion. The Instrument Society of America (ISA) has developed a complete list of symbols for instruments, instrument identification, and process control drawings, which will be discussed in Chap. 15.

The units of measurement fall into two distinct systems; first, the English system and second, the International system, SI (Systéme International D'Unités) based on the metric system, but there are some differences. The English system has been the standard used in the United States, but the SI system is slowly making inroads, so that students need to be aware of both systems of units and be able to convert units from one system to the other. Confusion can arise over some units such as pound mass and pound weight. The unit for pound mass is the slug (no longer in common use), which is the equivalent of the kilogram in the SI system of units whereas pound weight is a force similar to the newton, which is the unit of force in the SI system. The conversion factor of 1 lb = 0.454 kg, which is used to convert mass (weight) between the two systems, is in effect equating 1-lb force to 0.454-kg mass; this being the mass that will produce a force of 4.448 N or a force of 1 lb. Care must be taken not to mix units of the two systems. For consistency some units may have to be converted before they can be used in an equation.

Table 1.1 gives a list of the base units used in instrumentation and measurement in the English and SI systems and also the conversion factors, other units are derived from these base units.

Example 1.2 How many meters are there in 110 yard?

$$110 \text{ yard} = 330 \text{ ft} = (330 \times 0.305) \text{ m} = 100.65 \text{ m}$$

Example 1.3 What is the equivalent length in inches of 2.5 m?

$$2.5 \text{ m} = (2.5/0.305) \text{ ft} = 8.2 \text{ ft} = 98.4 \text{ in}$$

Example 1.4 The weight of an object is 2.5 lb. What is the equivalent force and mass in the SI system of units?

$$2.5 \text{ lb} = (2.5 \times 4.448) \text{ N} = 11.12 \text{ N}$$
$$2.5 \text{ lb} = (2.5 \times 0.454) \text{ kg} = 1.135 \text{ kg}$$

Table 1.2 gives a list of some commonly used units in the English and SI systems, conversion between units, and also their relation to the base units. As explained above the lb is used as both the unit of mass and the unit of force.

TABLE 1.1 Basic Units

Quantity	English		SI		
Base units	Units	Symbol	Units	Symbol	Conversion to SI
Length	Foot	ft	Meter	m	1 ft = 0.305 m
Mass	Pound (slug)	lb (slug)	Kilogram	kg	1 lb(slug) = 14.59 kg
Time	Second	s	Second	s	
Temperature	Rankine	R	Kelvin	K	1°R = 5/9 K
Electric current	Ampere	A	Ampere	A	

TABLE 1.2 Units in Common Use in the English and SI System

Quantity	English			SI		
	Name	Symbol	Units	Name	Symbol	Units
Frequency	Hertz			Hertz	Hz	s^{-1}
Energy	Foot-pound	ft·lb	$lb·ft^2/s^2$	Joule	J	$kg·m^2/s^2$
Force	Pound	lb	$lb·ft/s^2$	Newton	N	$kg·m/s^2$
Resistance	Ohm			Ohm	Ω	$kg·m^2$ per $(s^3·A^2)$
Electric Potential	Volt			Volt	V	$A·\Omega$
Pressure	Pound per in^2	psi	lb/in^2	Pascal	Pa	N/m^2
Charge	Coulomb			Coulomb	C	$A·s$
Inductance	Henry			Henry	H	$kg·m^2$ per $(s^2·A^2)$
Capacitance	Farad			Farad	F	$s^4·A^2$ per $(kg·m^2)$
Magnetic flux				Weber	Wb	$V·s$
Power	Horsepower	hp	$lb·ft^2/s^3$	Watt	W	J/s

Conversion to SI
1 ft·lb = 1.356 J
1 lb (F) = 4.448 N
1 psi = 6897 Pa
1 hp = 746 W

Hence, the unit for the lb in energy and power is mass, whereas the unit for the lb in pressure is force, where the lb (force) = lb (mass) × g (force due to gravity).

Example 1.5 What is the pressure equivalent of 18 psi in SI units?

$$1 \text{ psi} = 6.897 \text{ kPa}$$
$$18 \text{ psi} = (18 \times 6.897) \text{ kPa} = 124 \text{ kPa}$$

Standard prefixes are commonly used for multiple and submultiple quantities to cover the wide range of values used in measurement units. These are given in Table 1.3

1.6 Instrument Parameters

The *accuracy* of an instrument or device is the difference between the indicated value and the actual value. Accuracy is determined by comparing an indicated reading to that of a known standard. Standards can be calibrated devices or obtained from the National Institute of Standards and Technology (NIST).

TABLE 1.3 Standard Prefixes

Multiple	Prefix	Symbol	Multiple	Prefix	Symbol
10^{12}	tera	T	10^{-2}	centi	c
10^{9}	giga	G	10^{-3}	milli	m
10^{6}	mega	M	10^{-6}	micro	μ
10^{3}	kilo	k	10^{-9}	nano	n
10^{2}	hecto	h	10^{-12}	pico	p
10	deka	da	10^{-15}	femto	f
10^{-1}	deci	d	10^{-18}	atto	a

This is the government organization that is responsible for setting and maintaining standards, and developing new standards as new technology requires it. Accuracy depends on linearity, hysteresis, offset, drift, and sensitivity. The resulting discrepancy is stated as a ± deviation from the true value, and is normally specified as a percentage of full-scale reading or deflection (%FSD). Accuracy can also be expressed as the percentage of span, percentage of reading, or an absolute value.

Example 1.6 A pressure gauge ranges from 0 to 50 psi, the worst-case spread in readings is ±4.35 psi. What is the %FSD accuracy?

$$\%FSD = \pm (4.35 \text{ psi}/50 \text{ psi}) \times 100 = \pm 8.7$$

The *range* of an instrument specifies the lowest and highest readings it can measure, i.e., a thermometer whose scale goes from −40°C to 100°C has a range from −40°C to 100°C.

The *span* of an instrument is its range from the minimum to maximum scale value, i.e., a thermometer whose scale goes from −40°C to 100°C has a span of 140°C. When the accuracy is expressed as the percentage of span, it is the deviation from true expressed as a percentage of the span.

Reading accuracy is the deviation from true at the point the reading is being taken and is expressed as a percentage, i.e., if a deviation of ±4.35 psi in Example 1.6 was measured at 28.5 psi, the reading accuracy would be (4.35/28.5) × 100 = ±15.26% of reading.

Example 1.7 In the data sheet of a scale capable of weighing up to 200 lb, the accuracy is given as ±2.5 percent of a reading. What is the deviation at the 50 and 100 lb readings, and what is the %FSD accuracy?

$$\text{Deviation at 50 lb} = \pm (50 \times 2.5/100) \text{ lb} = \pm 1.25 \text{ lb}$$
$$\text{Deviation at 100 lb} = \pm (100 \times 2.5/100) \text{ lb} = \pm 2.5 \text{ lb}$$

Maximum deviation occurs at FSD, that is, ±5 lb or ±2.5% FSD

The *absolute accuracy* of an instrument is the deviation from true as a number not as a percentage, i.e., if a voltmeter has an absolute accuracy of ±3 V in the

100-volt range, the deviation is ±3 V at all the scale readings, e.g., 10 ± 3 V, 70 ± 3 V and so on.

Precision refers to the limits within which a signal can be read and may be somewhat subjective. In the analog instrument shown in Fig. 1.6a, the scale is graduated in divisions of 0.2 psi, the position of the needle could be estimated to within 0.02 psi, and hence, the precision of the instrument is 0.02 psi. With a digital scale the last digit may change in steps of 0.01 psi so that the precision is 0.01 psi.

Reproducibility is the ability of an instrument to repeatedly read the same signal over time, and give the same output under the same conditions. An instrument may not be accurate but can have good reproducibility, i.e., an instrument could read 20 psi as having a range from 17.5 to 17.6 psi over 20 readings.

Sensitivity is a measure of the change in the output of an instrument for a change in the measured variable, and is known as the transfer function, i.e., when the output of a pressure transducer changes by 3.2 mV for a change in pressure of 1 psi, the sensitivity is 3.2 mV/psi. High sensitivity in an instrument is preferred as this gives higher output amplitudes, but this may have to be weighted against linearity, range, and accuracy.

Offset is the reading of an instrument with zero input.

Drift is the change in the reading of an instrument of a fixed variable with time.

Hysteresis is the difference in readings obtained when an instrument approaches a signal from opposite directions, i.e., if an instrument reads a midscale value going from zero it can give a different reading from the value after making a full-scale reading. This is due to stresses induced into the material of the instrument by changing its shape in going from zero to full-scale deflection. Hysteresis is illustrated in Fig. 1.6b.

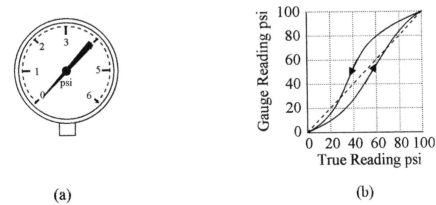

(a)   (b)

Figure 1.6 Gauges (a) pressure gauge showing graduations; (b) hysteresis curve for an instrument.

Example 1.8 A pressure gauge is being calibrated. The pressure is taken from 0 to 100 psi and back to 0 psi. The following readings were obtained on the gauge:

True Pressure (psi)	0	20	40	60	80	100	80	60	40	20	0
Gauge reading (psi)	1.2	19.5	37.0	57.3	81.0	104.2	83.0	63.2	43.1	22.5	1.5

Figure 1.7a shows the difference in the readings when they are taken from 0 going up to FSD and when they are taken from FSD going back down to 0. There is a difference between the readings of 6 psi or a difference of 6 percent of FSD, that is, ±3 percent from linear.

Resolution is the smallest amount of a variable that an instrument can resolve, i.e., the smallest change in a variable to which the instrument will respond.

Repeatability is a measure of the closeness of agreement between a number of readings (10 to12) taken consecutively of a variable, before the variable has time to change. The average reading is calculated and the spread in the value of the readings taken.

Linearity is a measure of the proportionality between the actual value of a variable being measured and the output of the instrument over its operating range. Figure 1.7b shows the pressure input versus voltage output curve for a pressure to voltage transducer with the best fit linear straight line. As can be seen, the actual curve is not a straight line. The maximum deviation of +5 psi from linear occurs at an output of 8 V and −5 psi at 3 V giving a deviation of ±5 psi or an error of ±5 percent of FSD.

The deviation from true for an instrument may be caused by one of the above or a combination of several of the above factors, and can determine the choice of instrument for a particular application.

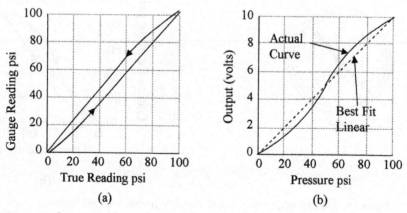

Figure 1.7 Instrument inaccuracies (a) hysteresis error of a pressure gauge; (b) non-linearity in a pressure-to-voltage transducer.

Summary

This chapter introduces the concept of process control and simple process loops, which will be expanded in later chapters.

The key points covered in this chapter are:

1. A description of the operation of a basic process loop with a definition of the terms used in process control
2. Some of the basic considerations for electrical, air, and water requirements in a process facility. Consideration needs for safety
3. A comparison of the units used for parameter measurement and their relation to the basic units
4. The relation between the English and the SI units, which are based on metric units. The use of standard prefixes to define multiples
5. The accuracy of sensors and instruments and parameters such as linearity, resolution, sensitivity, hysteresis, and repeatability, used to evaluate accuracy

Problems

1.1 What is the difference between controlled and manipulated variables?

1.2 What is the difference between set point, error signal, and correction signal?

1.3 How many pounds are equivalent to 63 kg?

1.4 How many micrometers are equivalent to 0.73 milli-in?

1.5 How many pounds per square inch are equivalent to 38.2 kPa?

1.6 How many foot-pounds of energy are equivalent to 195 J?

1.7 What force in pounds is equivalent to 385 N?

1.8 How many amperes are required from a 110-V supply to generate 1.2 hp? Assume 93- percent efficiency.

1.9 How many joules are equivalent to 27 ft·lb of energy?

1.10 What is the sensitivity of an instrument whose output is 17.5 mV for an input change of 7°C?

1.11 A temperature sensor has a range of 0 to 120°C and an absolute accuracy of ±3°C. What is its FSD percent accuracy?

1.12 A flow sensor has a range of 0 to 25 m/s and a FSD accuracy of ±4.5 percent. What is the absolute accuracy?

1.13 A pressure sensor has a range of 30 to 125 kPa and the absolute accuracy is ±2 kPa. What is its percent full-scale and span accuracy?

1.14 A temperature instrument has a range −20°F to 500°F. What is the error at 220°F? Assume the accuracy is (a) ±7 percent of FSD and (b) ±7 percent of span.

1.15 A spring balance has a span of 10 to 120 kg and the absolute accuracy is ±3 kg. What is its %FSD accuracy and span accuracy?

1.16 A digital thermometer with a temperate range of 129.9°C has an accuracy specification of ±1/2 of the least significant bit. What is its absolute accuracy, %FSD accuracy, and its resolution?

1.17 A flow instrument has an accuracy of (a) ±0.5 percent of reading and (b) 0.5%FSD. If the range of the instrument is 10 to 100 fps, what is the absolute accuracy at 45 fps?

1.18 A pressure gauge has a span of 50 to 150 psi and its absolute accuracy is ±5 psi. What is its %FSD and span accuracy?

1.19 Plot a graph of the following readings for a pressure sensor to determine if there is hysteresis, and if so, what is the hysteresis as a percentage of FSD?

True pressure (kPa)	0	20	40	60	80	100	80	60	40	20	0
Gauge pressure (kPa)	0	15	32	49.5	69	92	87	62	44	24	3

1.20 Plot a graph of the following readings for a temperature sensor to determine the linearity of the sensor. What is the nonlinearity as a percentage of FSD?

True pressure (kPa)	0	20	40	60	80	100
Gauge reading (kPa)	0	16	34	56	82	110

Chapter 2

Basic Electrical Components

Chapter Objectives

This chapter will help to refresh and expand your understanding of basic electrical components and the basic terms used in electricity as required for instrumentation.

This chapter discusses

- Basic passive components (resistors, capacitors, and inductors) used in electrical circuits
- Applications of Ohm's law and Kirchoff's laws
- Use of resistors as voltage dividers
- Effective equivalent circuits for basic devices connected in series and parallel
- The Wheatstone bridge
- Loading of instruments on sensing circuits
- Impedances of capacitors and inductors

It is assumed that the student has a basic knowledge of electricity and electronics and is familiar with basic definitions. To recap, the three basic passive components—resistors, capacitors, and inductors—as well as some basic formulas as applied to direct and alternating currents will be discussed in this section.

2.1 Introduction

Electrical power can be in the form of either direct current (dc) (one direction only) or alternating current (ac) (the current reverses periodically, see Fig. 2.1). In ac circuits the electromotive force drives the current in one direction then reverses itself and drives the current in the reverse direction. The rate of direction change is expressed as a frequency f and is measured in hertz (Hz), i.e., cycles per second.

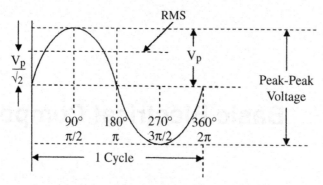

Figure 2.1 The basic sine wave.

Electrical signals travel at the speed of light. The distance traversed in one cycle is called a wavelength λ, the relationship between frequency and wavelength (meters) is given by the following equation:

$$f = \frac{c}{\lambda} \tag{2.1}$$

where c is the speed of light (3×10^8 m/s).

In both dc and ac circuits, conventional current was originally considered to flow from the more positive to the less positive or negative terminal. It was later discovered that current flow is really a flow of electrons (negative particles) that flow from negative to positive. To avoid confusion, only conventional current flow will be considered in this text, i.e., current flows from positive to negative. When measuring ac voltages and currents with a meter, the root mean square (rms) value is displayed. The rms value of a sine wave has the same effective energy as the dc value. When displaying sine waves on an oscilloscope it is often more convenient to measure the peak-to-peak (pp) values as shown in Fig. 2.1. The peak amplitude of the sine wave (V_p or I_p) (0 to peak) is then $(p-p)/2$, and the rms value is given by

$$\text{rms} = \frac{V_p}{\sqrt{2}} \tag{2.2}$$

The basic sine wave shown in Fig. 2.1 can be equated to a 360° circle or a circle with 2π rad. The period (cycle time) of a sine wave is broken down into four phases each being 90° or $\pi/2$ rad. This is derived from the trigonometry functions, and will not be elaborated upon.

2.2 Resistance

It is assumed that the student is familiar with the terms insulators, conductors, semiconductors, electrical resistance, capacitance, and inductance. Hence, the basic equations commonly used in electricity will be considered as a starting point.

2.2.1 Resistor formulas

The resistivity ρ of a material is the resistance to current flow between the opposite faces of a unit cube of the material (ohm per unit length). The resistance R of a component is expressed by

$$R = \frac{\rho l}{A} \tag{2.3}$$

where l is the length of the material (distance between contacts), and A is the cross-sectional area of the resistor; l and A must be in compatible units.

Table 2.1 gives the resistivity of some common materials. The resistivity ρ is temperature dependant, usually having a positive temperature coefficient (resistance increases as temperature increases), except for some metal oxides and semiconductors which have a negative temperature coefficient. The metal oxides are used for thermistors. The variation of resistance with temperature is given by

$$R_{T2} = R_{T1}(1 + \alpha T) \tag{2.4}$$

where R_{T2} = resistance at temperature T_2
R_{T1} = resistance at temperature T_1
α = temperature coefficient of resistance
T = temperature difference between T_1 and T_2

The variation of resistance with temperature in some materials (platinum) is linear over a wide temperature range. Hence, platinum resistors are often used as temperature sensors. See Example 8.10 in Chap. 8.

Ohm's law applies to both dc and ac circuits, and states that in an electrical circuit the electromotive force (emf) will cause a current I to flow in a resistance R, such that the emf is equal to the current times the resistance, i.e.

$$E = IR \tag{2.5}$$

This can also be written as

$$I = E/R \quad \text{or} \quad R = E/I$$

TABLE 2.1 **Resistivity of Some Common Materials**

Material	Resistivity (ohms per unit length)	Material	Resistivity (ohms per unit length)
Aluminum	17	Brass	42
Bronze	108	Chromel	420–660
Copper	10.4	German silver	200
Gold	14.6	Graphite	4800
Iron pure	59	Lead	132
Mercury	575	Nickel	42
Nichrome	550–660	Platinum	60
Silver	9.6	Steels	72–500
Tungsten	33		

where E = electromotive force in volts (V)
I = current in amperes (A)
R = resistance in ohms (Ω)

Example 2.1 The emf across a 4.7-kΩ resistor is 9 V. How much current is flowing?

$$I = \frac{E}{R} = \frac{9}{4.7 \times 10^3} \text{ A} = 1.9 \times 10^{-3} \text{ A} = 1.9 \text{ mA}$$

Power dissipation P occurs in a circuit, whenever current flows through a resistance. The power produced in a dc or ac circuit is given by

$$P = EI \tag{2.6}$$

where P is power in watts. (In ac circuits E and I are rms values).
Substituting Eq. (2.1) in Eq. (2.6) we get

$$P = I^2 R = \frac{E^2}{R} \tag{2.7}$$

In an ac circuit the power dissipation can also be given by

$$P = E_p I_p / 2 \tag{2.8}$$

where E_p and I_p are the peak voltage and current values.

Example 2.2 What is the dissipation in the resistor in Example 2.1?

$$P = EI = (9 \times 1.9) \text{ mW} = 17.1 \text{ mW}$$

Carbon composition resistors are available in values from 1 Ω to many megaohms in steps of 1, 2, 5, and 10 percent, where the steps are also the tolerances, as well as being available in different wattage ratings from 1/8 to 2 W. The wattage rating can be extended by using metal film or wire-wound resistors to several tens of watts. When choosing resistors for an application, not only should the resistor value be specified but the tolerance and wattage should also be specified. The value of carbon resistors is indicated by color bands and can be found in resistor color code charts.

Power transmission is more efficient over high-voltage lines at low current than at lower voltages and higher currents.

Example 2.3 Compare the energy loss of transmitting 5000 W of electrical power over power lines with an electrical resistance of 10 Ω using a supply voltage of 5000 V and the loss of transmitting the same power using a supply voltage of 1000 V through the same power lines.

The loss using 5000 V can be calculated as follows:

$$I = \frac{P}{E} = \frac{5000}{5000} \text{ A} = 1 \text{ A}$$

$$\text{Loss} = I^2 R = (1 \times 1 \times 10) \text{ W} = 10 \text{ W}$$

If, however, the supply voltage was 1000 V the loss would be

$$I = \frac{P}{E} = \frac{5000}{1000} \text{ A} = 5 \text{ A}$$

$$\text{Loss} = I^2 R = (5 \times 5 \times 10) \text{ W} = 250 \text{ W}$$

So that in going from 5000 to 1000 V, the losses increase from 10 to 250 W

2.2.2 Resistor combinations

Resistors can be connected in series, parallel, or a combination of both in a resistor network.

Resistors in series are connected as shown in Fig. 2.2a, their effective total value R_T is the sum of the individual resistors, and is given by

$$R_T = R_1 + R_2 + R_3 + \cdots + R_n \tag{2.9}$$

Example 2.4 What is the current flowing in the resistor network shown in Fig. 2.2a?

$$R_T = 12 \text{ k}\Omega + 5 \text{ k}\Omega + 24 \text{ k}\Omega = 41 \text{ k}\Omega$$

$$I_s = \frac{E}{R_T} = \frac{10}{41 \times 10^3} \text{ A} = 0.244 \text{ mA}$$

Voltage dividers are constructed using resistors connected in series as in Fig. 2.2a. A divider is used to reduce the supply voltage to a lower voltage value. The output voltage from the resistive divider can be calculated by multiplying the value of the current flowing by the value of the resistor across which the voltage is being measured, or by using the resistor ratios.

Example 2.5 What is the value of V_{out} across R_3 with respect to the negative battery terminal in Fig. 2.2a?

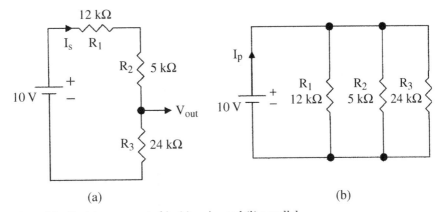

Figure 2.2 Resistors connected in (a) series and (b) parallel.

Since the current flowing is the same in all resistors

$$V_{out} = 0.244 \times 24 \text{ k}\Omega = 5.8 \text{ V}$$

Thus, using the resistance values in the example 5.86 V is obtained from a 10-V supply. Alternatively, V_{out} can be calculated as follows

$$I_s = \frac{E}{R_T} = \frac{V_{out}}{R_3}$$

From which we get

$$V_{out} = \frac{ER_3}{R_1 + R_2 + R_3} \quad (2.10)$$

This shows that the value of V_{out} is the supply voltage times the resistor ratios. Using this equation in Example 2.5

$$V_{out} = \frac{10 \times 24}{12 + 5 + 24} \text{ V} = 5.8 \text{ V}$$

Potentiometers are variable resistance devices that can be used to set voltages. They can have linear or logarithmic characteristics and can be constructed using carbon film tracks, or wire wound if longevity and accuracy is required (see Fig. 2.3b and c). A wiper or slider can traverse the track to give a variable voltage. A potentiometer is connected between a supply voltage and ground as shown in Fig. 2.3a. Using a linear potentiometer the wiper can be used to obtain a voltage proportional to its position on the track making a voltage divider. In Fig. 2.3b the output voltage is proportional to shaft rotation, and in Fig. 2.3c the output voltage is proportional to linear displacement. Linear potentiometers are used to convert mechanical movement into electrical voltages. Logarithmic devices are used in volume controls (the ear, for instance, has a logarithmic response) or similar applications, where a logarithmic output is required.

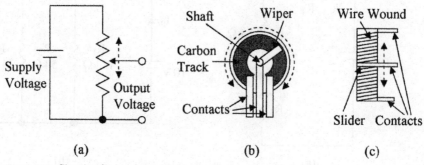

Figure 2.3 Circuit of (a) voltage divider potentiometer, (b) rotational carbon potentiometer, and (c) wire-wound slider type potentiometer.

Resistors in parallel are connected as shown in Fig. 2.2b, and their total effective value R_T is given by

$$\frac{1}{R_T} = \frac{1}{R_1} + \frac{1}{R_2} + \frac{1}{R_3} + \cdots + \frac{1}{R_n} \qquad (2.11)$$

Example 2.6 What is the current I_p flowing in the circuit shown in Fig. 2.2b, and what is the equivalent value R_T of the three parallel resistors?

$$\frac{1}{R_T} = \frac{1}{12 \text{ k}\Omega} + \frac{1}{5 \text{ k}\Omega} + \frac{1}{24 \text{ k}\Omega} = \frac{5 \times 24 + 12 \times 24 + 5 \times 12}{120 \text{ k}\Omega} = \frac{39 \text{ k}\Omega^{-1}}{120}$$

$$R_T = 120 \text{ k}\Omega/39 = 3.08 \text{ k}\Omega$$

$$I_p = 10/3.08 \text{ k}\Omega = 3.25 \text{ mA}$$

Kirchoff's laws apply to both dc and ac circuits. The fist law (voltage law) states that in any closed path in a circuit, the algebraic sum of the voltages is zero, or the sum of the voltage drops across each component in a series circuit is equal to the source voltage. From Fig. 2.4a we get

$$-E + V_1 + V_2 + V_3 = 0 \quad \text{or} \quad E = V_1 + V_2 + V_3 \qquad (2.12)$$

Kirchoff's second law (current law) states that the sum of the currents at any node or junction is zero, i.e., the current flowing into a node is equal to the current flowing out of the node. In Fig. 2.4b for the upper node we get

$$-I_T + I_1 + I_2 + I_3 = 0 \quad \text{or} \quad I_T = I_1 + I_2 + I_3 \qquad (2.13)$$

The Wheatstone bridge is the most common resistance network developed to measure small changes in resistance and is often used in instrumentation with resistive types of sensors. The bridge circuit is shown in Fig. 2.5a. Four resistors are connected in the form of a diamond with the supply and measuring instrument forming the diagonals. When all the resistors are equal the bridge

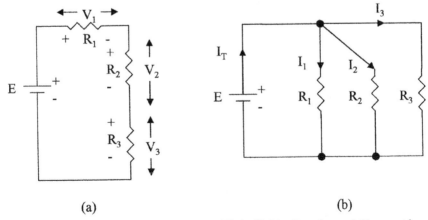

Figure 2.4 Resistor networks to demonstrate Kirchoff's (a) voltage law and (b) current law.

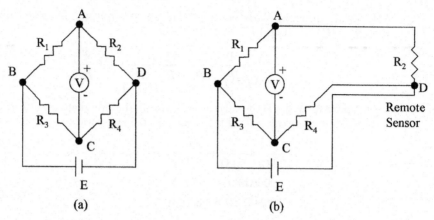

Figure 2.5 Circuit of (a) Wheatstone bridge and (b) compensation for lead resistance used in remote sensing.

is balanced, i.e., the voltage at A and C are equal ($E/2$) and the voltmeter reads zero.

If R_2 is the resistance of a sensor whose change in value is being measured, the voltage at A will increase with respect to C as the resistance value increases, so that the voltmeter will have a positive reading. The voltage will change in proportion to any changes in the value of R_2, making the bridge very sensitive to small changes in resistance. A bridge circuit can also be used to compensate for changes in resistance due to temperature changes, i.e., if R_1 and R_2 are the same type of sensing element, such as a strain gauge and reference strain gauge (see Fig. 2.6). The resistance of each gauge will change by an equal percentage with temperature, so that the bridge will remain balanced when the temperature changes. If R_2 is now used to sense a variable,

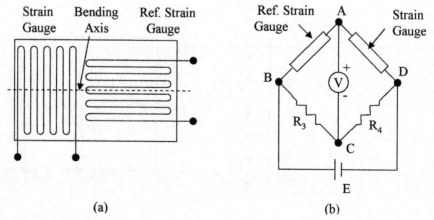

Figure 2.6 Showing (a) strain gauge with reference gauge and (b) strain gauges used in a Wheatstone bridge.

the voltmeter will only sense the change in R_2 due to the change in the variable, as the effects of temperature changes on R_1 and R_2 will cancel.

Because of the above two features, bridges are extensively used in instrumentation. The voltmeter (measuring instrument) should have a high resistance, so that it does not load the bridge circuit. Bridges can also be used with ac supply voltages and ac meters. The resistors can then be replaced with capacitors, inductors, or a combination of resistors, capacitors, and inductors.

In many applications, the sensing resistor (R_2) can be remote from a centrally located bridge. In such cases the resistance of the leads can be zeroed out by adjusting the bridge resistors. Any change in lead resistance due to temperature, however, will appear as a sensor value change. To correct for this error, lead compensation can be used. This is achieved by using three interconnecting leads as shown in Fig. 2.5b. A separate power lead is used to supply R_2 so that only signal current flows in the signal lead from R_2 to the bridge resistor R_4. Any variations in voltage drop due to the supply current in the lead resistance do not affect the balance of the bridge. However, by monitoring any voltage changes between R_4 and the voltage at the negative battery terminal a correction voltage that can be applied to the lead between R_2 and R_1 can be obtained, and this lead will also carry the supply current back to the bridge, and any changes in lead resistance will affect both leads equally.

Example 2.7 The resistors in the bridge circuit shown in Fig. 2.5a are all 2.7 kΩ, except R_1 which is 2.2 kΩ. If $E = 15$ V what will the voltmeter read?

The voltage at point C will be 7.5 V, as $R_3 = R_4$, the voltage at $C = \frac{1}{2}$ the supply voltage. The voltage at A will be given by

$$E_{AD} = \frac{E \times R_2}{R_1 + R_2} = \frac{15 \text{ V} \times 2.7 \text{ k}\Omega}{2.2 \text{ k}\Omega + 2.7 \text{ k}\Omega} = \frac{40.5 \text{ V}}{4.9} = 8.26 \text{ V}$$

The voltmeter will read 8.26 − 7.5 V = 0.76 V (note meter polarity)

2.2.3 Resistive sensors

Strain gauges are examples of resistive sensors (see Fig. 2.6a). The resistive conducting path in the gauge is copper or nickel particles deposited onto a flexible substrate in a serpentine form. When the substrate is bent in a concave shape along the bending axis perpendicular to the direction of the deposited resistor, the particles are compressed and the resistance decreases. If the substrate is bent in the other direction along the bending axis, the particles tend to separate and the resistance increases. Bending along an axis perpendicular to the bending axis does not compress or separate the particles in the strain gauge; so the resistance does not change. Piezoresistors are also used as strain gauge elements. These devices are made from certain crystalline materials such as silicon. The material changes its resistance when strained similarly to the deposited strain gauge. These devices can be very small. The resistance change in strain gauge elements is proportional to the degree of bending, i.e., if the gauge was attached to a pressure sensing diaphragm and pressure is applied to one side of the diaphragm, the diaphragm bows in relation to the pressure applied. The change in resistance of the strain gauge attached to the diaphragm is then proportional to the pressure applied. Figure 2.6b shows a Wheatstone bridge connected to the strain gauge elements of a pressure sensor. Because the resistance of the strain gauge element is temperature-sensitive, a reference strain gauge is also added to the bridge to compensate for these changes. This second strain gauge is positioned adjacent to the first so that it is at the same temperature, but rotated 90°, so that it is at right angles to the pressure-sensing strain gauge element and will, therefore, not sense the deformation as seen by the pressure-sensing element.

2.3 Capacitance

2.3.1 Capacitor formulas

Capacitors store electrical charge, as opposed to cells where the charge is generated by chemical action. Capacitance is a measure of the amount of charge that can be stored. The capacitance of a capacitor is given by

$$C = \varepsilon A/d \tag{2.14}$$

where C = capacitance in farads (F)
ε = dielectric constant of the material (F/m) between the plates
A = area of the plates (m^2)
d = distance between the plates (m)

The dielectric constants of some common materials are given in Table 2.2. A 1-F capacitor is defined as a capacitor that will store 1 C of charge when there is a voltage potential of 1 V across the plates of the capacitor (a coulomb of charge is obtained when a current of 1 A flows for 1 s). A farad is a very large unit and microfarad and picofarad are the commonly used units.

Example 2.8 What is the capacitance between two parallel plates whose areas are 1 m^2 separated by a 1-mm thick piece of dielectric with a dielectric constant of 5.5×10^{-9} F/m?

$$C = \frac{\varepsilon A}{d} = \frac{5.5 \times 10^{-9} \text{ F/m} \times 1 \text{ m}^2}{1 \times 10^{-3} \text{ m}}$$

$$= 5.5 \times 10^{-6} \text{ F} = 5.5 \text{ µF}$$

In electrical circuits, capacitors are used to block dc voltages, but will allow ac voltages to pass through them. Capacitors do, however, present impedance not resistance to ac current flow. This is due to the fact that the current and

TABLE 2.2 Dielectric Constants of Some Common Materials

Material	Dielectric constant × 10^{-9} F/m	Material	Dielectric constant × 10^{-9} F/m
Glass	5–10	Plexiglas	3.4
Mica	3–6	Polyethylene	2.35
Mylar	3.1	Polyvinyl chloride	3.18
Neoprene	6.7	Teflon	2.1
Germanium	16	Paper	2

voltage are not in phase. Impedance is similar to the resistance a resistor presents to a dc current flow, but as they are not identical they cannot be directly added and will be dealt with in Chap. 3.

The impedance of a capacitor to ac flow is given by

$$X_C = \frac{1}{2\pi f C} \qquad (2.15)$$

where X_C = impedance to ac current flow
f = frequency of the ac signal
C = capacitance in farads

Ohm's law also applies to ac circuits, so that the relation between voltage and current is given by

$$E = IX_C \qquad (2.16)$$

where E is the ac voltage amplitude and I is the ac current flowing.

Example 2.9 What is the ac current flowing in the circuit shown in Fig. 2.7a?

$$X_C = \frac{1}{2\pi f C} = \frac{1}{2 \times 3.142 \times 1500 \times 0.1 \times 10^{-6}} \; \Omega$$

$$= 1.06 \times 10^3 \; \Omega = 1.06 \; k\Omega$$

$$I = E/X_C = 12/1.06 \times 10^3 = 11.3 \times 10^{-3} \; A = 11.3 \; mA$$

2.3.2 Capacitor combinations

The formulas for the effective capacitance of capacitors connected in series and parallel are the opposite of resistors connected in series and parallel.

Capacitors in series are shown in Fig. 2.7b and have an effective capacitance given by

$$\frac{1}{C_T} = \frac{1}{C_1} + \frac{1}{C_2} + \frac{1}{C_3} + \cdots + \frac{1}{C_n} \qquad (2.17)$$

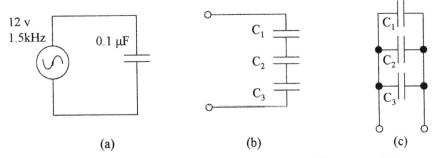

Figure 2.7 Circuits (a) used in Example 2.9 (b) capacitors connected in series, and (c) capacitors connected in parallel.

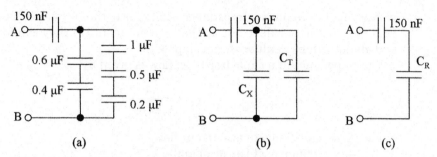

Figure 2.8 Circuits for use in Example 2.10: (*a*) circuit used in example, (*b*) reduction of series capacitors to single capacitors, and (*c*) replacement of parallel capacitors with a single capacitor.

Capacitors in parallel are shown in Fig. 2.7c and have an effective capacitance given by

$$C_T = C_1 + C_2 + C_3 + \cdots + C_n \tag{2.18}$$

Example 2.10 What is the value of the capacitor that could be used to replace the capacitor network shown in Fig. 2.8a?

In this example the first step is to reduce the three capacitors in series to a single capacitor and the two capacitors in series to a single capacitor as shown in Fig. 2.8b.

$$\frac{1}{C_T} = \frac{1}{1\ \mu F} + \frac{1}{0.5\ \mu F} + \frac{1}{0.2\ \mu F} = (1 + 2 + 5)\ \mu F^{-1}$$

$$C_T = 0.125\ \mu F$$

$$\frac{1}{C_X} = \frac{1}{0.6\ \mu F} + \frac{1}{0.4\ \mu F}$$

$$C_X = 0.24\ \mu F$$

The two capacitors in parallel in Fig. 2.8b are given by

$$C_R = C_X + C_T = 0.125\ \mu F + 0.24\ \mu F = 0.365\ \mu F$$

From Fig. 2.8c, the equivalent capacitance C_E is given by

$$\frac{1}{C_E} = \frac{1}{C_R} + \frac{1}{150\ nF} = \frac{1}{0.365\ \mu F} + \frac{1}{150\ nF} = \frac{1}{365\ nF} + \frac{1}{150\ nF}$$

$$C_E = \frac{54.75\ nF}{0.515} = 106.3\ nF$$

2.4 Inductance

2.4.1 Inductor formulas

Inductors are devices that oppose any change in the current flowing through them. The inductance of a coil is given by

$$L = \frac{N^2 \mu A}{d} \tag{2.19}$$

where L = inductance in henries
N = number of turns of wire
μ = permeability of the core of the coil (H/m)
A = cross sectional area of the coil (m^2)
d = length of the coil (m)

A henry is defined as the inductance that will produce an emf of 1 V when the current through the inductance changes at the rate of 1 A/s.

Example 2.11 A coil with a diameter of 0.5 m and length 0.7 m is wound with 100 turns of wire, what is its inductance if the material of the core has a permeability of 7.5×10^{-7} H/m?

$$L = \frac{100^2 \times 7.5 \times 10^{-7} \times 3.142 \times 0.5^2 \text{ H}}{4 \times 0.7} = \frac{5.85 \times 10^{-3} \text{ H}}{2.8} = 2.1 \times 10^{-3} \text{ H} = 2.1 \text{ mH}$$

Inductive impedance to ac current flow is given by

$$X_L = 2\pi f L \tag{2.20}$$

where X_L = impedance to ac current flow
f = frequency of the ac signal
L = inductance in henries

Example 2.12 What is the impedance to a 50-kHz sine wave of a 10-mH inductance?

$$X_L = 2\pi f L = 2\pi \times 50 \times 10^3 \times 10 \times 10^{-3} = 3100 \text{ } \Omega = 3.1 \text{ k}\Omega$$

2.4.2 Inductor combinations

The formula for the effective inductance of inductors connected in series and parallel is the same as for resistors.

Inductors in series have an effective inductance given by

$$L_T = L_1 + L_2 + L_3 + \cdots + L_n \tag{2.21}$$

Inductors in parallel have an effective inductance given by

$$\frac{1}{L_T} = \frac{1}{L_1} + \frac{1}{L_2} + \frac{1}{L_3} + \cdots + \frac{1}{L_n} \tag{2.22}$$

Summary

This chapter was designed to refresh and expand your knowledge of basic electrical components. The main points covered in this chapter are:

1. Introduction to the different effects of dc and ac electrical supplies on circuit components
2. Resistivity of materials and their resistance when made into components, the effect of temperature on the resistance of components, introduction to Ohm's law, and power dissipation in resistive components

3. The effective resistance of resistors connected in series and parallel and their use as voltage dividers
4. Discussion of Kirchoff's voltage and current laws, Wheatstone bridge circuits and their use in the measurement of small changes in resistance, and the use of bridge circuits for strain gauge measurement
5. Description of capacitance and the formulas used for capacitors, the effective capacitance of capacitors connected in series and parallel and the impedance of capacitors when used in ac circuits
6. A description of inductance and the formulas used for inductors, the effective impedance of inductors used in ac circuits, and the effective inductance of inductors when they are connected in series and parallel

Problems

2.1 A radio beacon transmits a frequency of 230 MHz. What is the wavelength of the signals?

2.2 What is the power dissipation in a 68 Ω resistive load, when a 110-V (peak-to-peak) sine wave is applied to the resistor?

2.3 The resistivity of a material used to make a round 950 Ω resistor is 53 Ω per unit length. If the resistor has a radius of 0.16 in, what is it's length?

2.4 A resistor with a temperature coefficient of 0.0045/°C has a resistance of 130 Ω at 20°C. At what temperature will the resistance be 183 Ω?

2.5 A dc voltage of 17 V is measured across a 133-Ω resistor. What is the current flowing through the resistor?

2.6 A dc voltage is applied to three resistors in parallel. The values of the resistors are 7.5, 12.5, and 14.8 kΩ. If the total current flowing is 2.7 mA, what is the applied voltage?

2.7 The configuration of the three resistors in Prob. 2.6 is changed from a parallel to a series connection. If the current flowing in the resistors is unchanged, what is the total voltage across the three resistors?

2.8 What is the supply current I_t flowing in the circuit shown in Fig. 2.9a?

2.9 Calculate the voltage across each of the resistors in Prob. 2.7. Does the result support Kirchoff's first law?

2.10 What is the current flowing in each of the resistors in Prob. 2.6? Does the result support Kirchoff's second law?

2.11 What is the voltage measured in the bridge circuit shown in Fig. 2.9b?

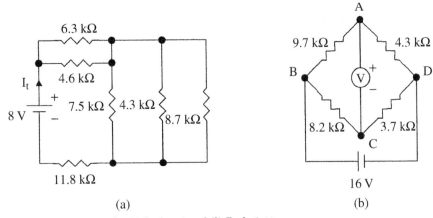

Figure 2.9 Diagrams for (a) Prob. 2.8 and (b) Prob. 2.11.

2.12 Two rectangular parallel plates 2.2 m by 3.7 m are separated by a material with a dielectric constant of 4.8×10^{-9} F/m. If the capacitance between the plates is 4.3 µF, what is the separation of the plates?

2.13 A 3.2 nF capacitor has an impedance of 0.02 MΩ when an ac voltage is applied to it. What is the frequency of the ac voltage?

2.14 What is the current flowing in Prob. 2.13, if the peak-to-peak ac voltage is 18 V?

2.15 Three capacitors are connected in series. See Fig. 2.7b. If the values of the capacitors are 110, 93, and 213 pF, what is the value of a single capacitor that could be used to replace them?

2.16 What is the value of a single capacitor that could be used to replace the capacitors shown in Fig. 2.10a?

2.17 An inductor of 2.8 mH is being constructed on a core whose diameter is 1.4 cm and length is 5.6 cm. If the permeability of the core is 4.7×10^{-7} H/m, how many turns of wire will be required?

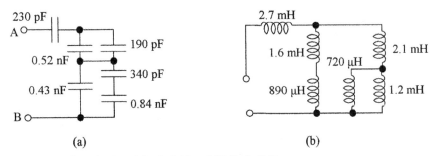

Figure 2.10 Circuits for (a) Prob. 2.16 and (b) Prob. 2.20.

2.18 What is the value of inductance that will have an impedance of 11.4 kΩ at a frequency of 2.3 MHz?

2.19 What value of inductance can be used to replace two inductors connected in parallel, if their values are 4.2 and 8.7 mH?

2.20 What value of inductance would be used to replace the inductor network shown in Fig. 2.10*b*?

Chapter 3

AC Electricity

Chapter Objectives

This chapter deals with the basic passive components in ac circuits, and will help you become familiar with these devices and understand their application in instrumentation.

This chapter discusses the following:

- Use of impedance as opposed to resistance in ac circuits
- Equivalent circuit of a combination of components in ac circuits
- Effective impedance to current flow of components in ac circuits
- Time delays and time constants
- Concept of phase angle
- Resonant frequency
- Concept of filters
- AC Wheatstone bridge
- Magnetic fields, meters, and motors
- Transformers

3.1 Introduction

Three basic components, resistors, capacitors, and inductors, are very important elements in electrical circuits as individual devices, or together. Resistors are used as loads, delays, and current limiting devices. Capacitors are used as dc blocking devices, in level shifting, integrating, differentiating, frequency determination, selection, and delay circuits. Inductors are used for frequency selection and ac blocking, in analog meter movements and relays, and are the basis for transformers and motors.

3.2 Circuits with R, L, and C

3.2.1 Voltage step

When a dc voltage is applied to a capacitor through a resistor, a current flows charging the capacitor (see Fig. 3.1a). Initially, all the voltage drops across the resistor; although current is flowing into the capacitor, there is no voltage drop across the capacitor. As the capacitor charges, the voltage across the capacitor builds up on an exponential, and the voltage across the resistor starts to decline, until eventually the capacitor is fully charged and current ceases to flow. The voltage across the capacitor is then equal to the supply voltage and the voltage across the resistor is zero. This is shown in Fig. 3.1b.

Two effects should be noted. The first is that the current flowing through the resistor and into the capacitor is the same for both components, but the voltages across each component is different, i.e., when the current flowing through the resistor is a maximum, the voltage across the resistor is maximum, given by $E = IR$, and the voltage is said to be in phase with the current. But in the case of the capacitor the voltage is zero when the current flowing is a maximum, and the voltage is a maximum when the current is zero. In this case the voltage lags the current or there is a phase shift between the voltage and the current of 90°. The second effect is that the voltage across the capacitor builds up at an exponential rate that is determined by the value of the resistor and the capacitor.

Similarly, if a dc voltage is applied to an inductance via a resistance as shown in Fig. 3.2a, the inductance will initially appear as a high impedance preventing current from flowing, so that the current will be zero, the supply voltage will appear across the inductance, and there will be zero voltage across the resistor. After the initial turn-on, current will start to flow and build up. The voltage across the resistor increases and starts to decrease across the inductance allowing the current to build up exponentially, until the current flow is limited by the resistance at its maximum value and the voltage across the inductance is zero. This is shown in Fig. 3.2b. The effects are similar in that the same current is flowing in

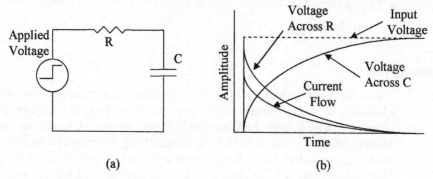

Figure 3.1 To demonstrate input transient: (a) circuit with resistance and capacitance and (b) associated waveforms.

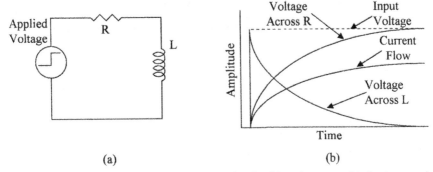

Figure 3.2 To demonstrate input transient (*a*) circuit with resistance and inductance and (*b*) associated waveforms.

both devices, the voltage and current in the resistor are in phase, but in the inductor are out of phase, i.e., in this case the voltage appears across the inductance before the current starts to flow, and goes to zero when the current is at its maximum, so that the voltage leads the current, and there is a phase shift between the voltage and the current of 90°. The voltage across the resistor increases at an exponential rate that is determined by the value of the inductance and resistance.

3.2.2 Time constants

When a step voltage is applied to an RC network in Fig. 3.1a, the voltage across the capacitor is given by the equation

$$E_C = E\left(1 - e^{-t/RC}\right) \tag{3.1}$$

where E_C = voltage across the capacitor at any instant of time
E = source voltage
t = time (seconds) after the step is applied

R is in ohms and C is in farads. If after the capacitor is fully charged the step input voltage is returned to zero, C will discharge and the voltage across the capacitor will be given by the equation

$$E_C = E e^{-t/RC} \tag{3.2}$$

Similar equations apply to the rise and fall of currents in an inductive circuit. These equations are, however, outside the scope of this course, and will not be taken further. They serve only to introduce circuit time constants.

The time constant of the voltage in a capacitive circuit from Eqs. (3.1) and (3.2) is defined as

$$t = CR \tag{3.3}$$

where t is the time (seconds) it takes for the voltage to reach 63.2 percent of its final or aiming voltage after the application of an input voltage step (charging or discharging), i.e., by the end of the first time constant the voltage across the capacitor will reach 6.32 V when a 10-V step is applied. During the second time constant the voltage across the capacitor will rise another 63.2 percent of the remaining voltage step, i.e., $(10 - 6.32) \text{ V} \times 63.2\% = 2.33$ V, or at the end of the two time constant periods, the voltage across the capacitor will be 8.65 V, and at the end of three periods 9.5 V, and so on, as shown in Fig. 3.3a. The voltage across the capacitor reaches 99 percent of its value in 5 CR.

Example 3.1 What is the time constant for the circuit shown in Fig. 3.1a if the resistor has a value of 220 kΩ and the capacitor is 2.2 µF?

$$t = 2.2 \times 3^{-6} \times 220 \times 10^3 \text{ s} = 484 \times 3^{-3} \text{ s} = 0.484 \text{ s}$$

The RC time constant is often used as the basis for time delays, i.e., a comparator circuit is set to detect when a voltage across a capacitor in a CR network reaches 63.2 percent of the input step voltage. The time delay generated is then 1 CR.

Capacitors can also be used for level shifting and signal integration. Figure 3.3b shows a 0 to 10-V step applied to a capacitor, and the resulting waveform. The 10 V step passes through the capacitor, but the output side of the capacitor is referenced by the resistor R to 10 V so that the step at V_{out} goes from 10 to 20 V, the voltage then decays back to 10 V in a time set by the CR time constant, i.e., the leading edge of the square wave has been level shifted by blocking the dc level of the input with the capacitor and applying a new dc level of 10 V. The decay of the square wave at the output is referred to as integration, i.e., a capacitor only lets a changing voltage through.

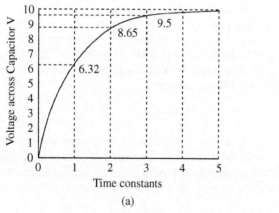
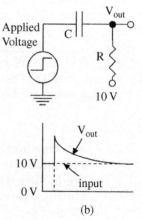

Figure 3.3 Shown is (a) plot of the voltage across a capacitor versus the circuit time constant and (b) an example of level shifting and integration using a capacitor.

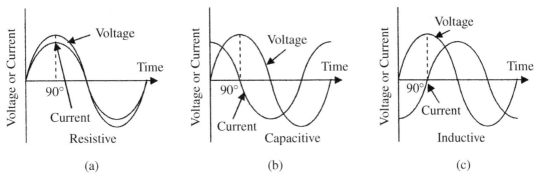

Figure 3.4 Phase relation in (a) resistive, (b) capacitive, and (c) inductive circuits.

In the case of an inductive circuit, the time constant for the current is given by

$$t = L/R \qquad (3.4)$$

where L is the inductance in henries, and t gives the time for the current to increase to 63.2 percent of its final current through the inductor.

3.2.3 Phase change

A phase change or shift that occurs between voltage and current in capacitors and inductors when a step voltage waveform is applied to them has been discussed. The same phase shift also takes place when an ac sine wave is applied to C, L, and R circuits, as shown in Fig. 3.4. In resistive elements (a) the current and voltage are in phase, in capacitive circuits (b) the current leads the voltage by 90° (Fig. 3.1), and in inductive circuits (c) the current lags the voltage again by 90° (Fig. 3.2).

Because the voltages and the currents are not in phase in capacitive and inductive ac circuits, these devices have impedance not resistance, and therefore, as already noted, impedance and resistance cannot be directly added. If a resistor, capacitor, and inductor are connected in series as shown in Fig. 3.5a,

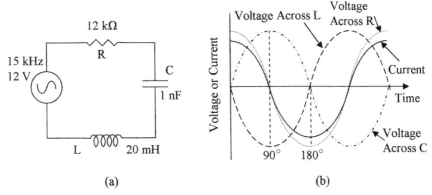

Figure 3.5 Shows (a) series R, C, and L circuit, and (b) waveforms and phase relations in a series circuit.

the same current will flow through all the three devices, but the voltages in the capacitor and inductor will be 180° out of phase and 90° out of phase with the voltage in the resistor, as shown in Fig. 3.5b. They can, however, be combined using vectors to give

$$\mathbf{E}^2 = \mathbf{V}_R^2 + (\mathbf{V}_L - \mathbf{V}_C)^2 \qquad (3.5)$$

where $\mathbf{E}$ = supply voltage
$\mathbf{V}_R$ = voltage across the resistor
$\mathbf{V}_L$ = voltage across the inductor
$\mathbf{V}_C$ = voltage across the capacitor

The vector addition of the voltages is shown in Fig. 3.6. In (a) the relations between $\mathbf{V}_R$, $\mathbf{V}_L$, and $\mathbf{V}_C$ are given; $\mathbf{V}_L$ and $\mathbf{V}_C$ lie on the x axis with one positive and the other negative because they are 180° out of phase; that is to say, they are of opposite sign, so that they can be subtracted to give the resulting $\mathbf{V}_C - \mathbf{V}_L$ vector; and $\mathbf{V}_R$ lies at right angles (90°) on the y axis. In (b) the $\mathbf{V}_C - \mathbf{V}_L$ vector and $\mathbf{V}_R$ vectors are shown with the resulting $\mathbf{E}$ vector, which from the trigonometry function gives Eq. (3.5).

The impedance Z of the circuit, as seen by the input is given by

$$Z = \sqrt{(R^2 + [X_L - X_C]^2)} \qquad (3.6)$$

where X_C and X_L are given by Eqs. (2.15) and (2.20).

The current flowing in the circuit can be obtained from Ohm's law, as follows:

$$I = \frac{E}{Z} \qquad (3.7)$$

Example 3.2 What is the current flowing in the circuit shown in Fig. 3.5a?

$$X_L = 2\pi f L = 2 \times 3.142 \times 15 \times 10^3 \times 20 \times 3^{-3} = 1.88 \text{ k}\Omega$$

$$X_C = \frac{1}{2\pi f C} = \frac{1}{2 \times 3.142 \times 15 \times 10^3 \times 1 \times 3^{-9}} \Omega = \frac{10^6}{94.26} \Omega = 10.6 \text{ k}\Omega$$

$$Z = \sqrt{(R^2 + [X_L - X_C])} = \sqrt{\{(12 \times 10^3)^2 + [1.88 \times 10^3 - 10.6 \times 10^3]^2\}}$$

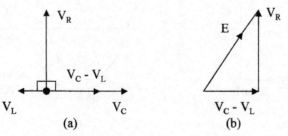

Figure 3.6 Shown are (a) the voltage vectors for the series circuit in Fig. 3.5 and (b) the resulting voltage $\mathbf{E}$ vector.

$$Z = \sqrt{\{144 \times 10^6 + [-8.72 \times 10^3]^2\}} = \sqrt{\{144 \times 10^6 + 76 \times 10^6\}}$$
$$Z = \sqrt{220 \times 10^6} = 14.8 \times 10^3 \; \Omega = 14.8 \; \text{k}\Omega$$
$$I = E/Z = 12/14.8 \times 10^3 = 0.81 \; \text{mA}$$

X_L and X_C are frequency dependant, and as the frequency increases X_L increases and X_C decreases. A frequency can be reached where X_L and X_C are equal, and the voltages across these components are equal and opposite, and cancel. At this frequency $Z = R$, $E = IR$, and the current is maximum. This frequency is called the resonant frequency of the circuit. At resonance

$$2\pi f L = \frac{1}{2\pi f C} \tag{3.8}$$

which can be rewritten for frequency as

$$f = \frac{1}{2\pi \sqrt{LC}} \; \text{Hz} \tag{3.9}$$

When the input frequency is below the resonant frequency, X_C is larger than X_L and the circuit is capacitive, and above the resonant frequency, X_L is larger than X_C and the circuit is inductive. Plotting the input current against the input frequency shows a peak in the input current at the resonant frequency, as shown in Fig. 3.7a.

Example 3.3 What is the resonant frequency of the series circuit in Fig. 3.5a? What is the current at this frequency?

Using Eq. 3.9 we get

$$f = \frac{1}{2\pi \sqrt{LC}} \; \text{Hz} = \frac{1}{2 \times 3.142 \times \sqrt{10^{-9} \times 20 \times 10^{-3}}} \; \text{Hz}$$

$$f = \frac{1}{2 \times 3.142 \times 4.47 \times 10^{-6}} \; \text{Hz} = \frac{10^6}{28.1} \; \text{Hz} = 3.56 \times 10^4 \; \text{Hz}$$

$$f = 35.6 \; \text{kHz}$$

The current can be obtained using Eq. (3.7):

$$I = E/Z = 12/12 \times 10^3 = 1 \; \text{mA}$$

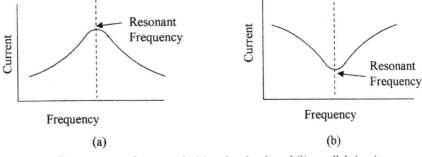

Figure 3.7 Current versus frequency in (a) series circuit and (b) parallel circuit.

38 Chapter Three

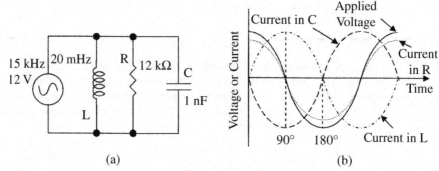

Figure 3.8 (a) A parallel R, C, and L circuit and (b) waveforms and phase relations in a parallel circuit.

In a parallel circuit as shown in Fig. 3.8a each component sees the same voltage but not the same current as is shown by the waveforms in Fig. 3.8b.

The source current I_S is the vector sum of the currents in each component, and is given by

$$I_S^2 = I_R^2 + (I_L - I_C)^2 \tag{3.10}$$

The impedance of the circuit Z as seen by the input is given by

$$\frac{1}{Z^2} = \frac{1}{R^2} + \frac{1}{(X_L - X_C)^2} \tag{3.11}$$

At the resonant frequency, I_L and I_C become equal and cancel so that $E = IR$. This can be seen from Eq. (3.10). Below the resonant frequency the circuit is inductive, and above the resonant frequency the circuit is capacitive. Plotting the current against frequency shows that the current is minimum at the resonant frequency, as shown in the frequency plot in Fig. 3.7b. The frequency at resonance is given by Eq. (3.9) and the current by Eq. (3.7).

3.3 RC Filters

Networks using resistors and capacitors are extensively used and sometimes small inductors are used in instrumentation circuits for filtering out noise, frequency selection, frequency rejection, and the like. Filters can be either passive or active (using amplifiers) and can be divided into the following:

High pass	Allows high frequencies to pass but blocks low frequencies.
Low pass	Allows low frequencies to pass but blocks high frequencies.
Band pass	Allows a specific range of frequencies to pass.
Band reject	Blocks a specific range of frequencies.

These passive filters are shown in Fig. 3.9. The number of resistive and capacitive elements determines whether the filter is a first-order filter, second-order

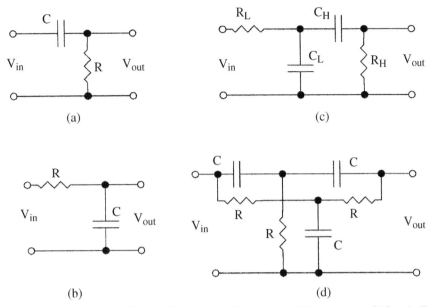

Figure 3.9 Examples of filters: (a) high pass, (b) low pass, (c) band pass, and (d) twin T band reject.

filter, and so on. The circuit configuration determines the characteristics of the filters, some of these classifications are Butterworth, Bessel, Chebyshev, and Legendre.

These are examples of the uses of resistors and capacitors in RC networks; further description is beyond the scope of this text.

3.4 AC Bridges

The concept of dc bridges described in Chap. 2 can also be applied to ac bridges. The resistive elements are replaced with impedances and the bridge supply is now an ac voltage, as shown in Fig. 3.10a. The differential voltage δV across S is then given by

$$\delta V = E \frac{Z_2 Z_3 - Z_1 Z_4}{(Z_1 + Z_3)(Z_2 + Z_4)} \qquad (3.12)$$

where E is the ac supply EMF.

When the bridge is balanced $\delta V = 0$ and Eq. (3.12) reduces to

$$Z_2 Z_3 = Z_1 Z_4 \qquad (3.13)$$

Example 3.4 What are the conditions for the bridge circuit in Fig. 3.10b to be balanced?

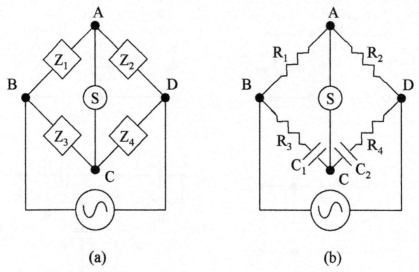

Figure 3.10 AC bridges (a) using block impedances and (b) bridge with R and C components for Example 3.4.

To be balanced Eq. (3.13) applies. There are two conditions that must be met for this equation to be balanced because of the phase shift produced by the capacitors. First, the resistive component must balance, and this gives

$$R_2 R_3 = R_1 R_4 \qquad (3.14)$$

Second, the impedance component must balance, and this gives

$$C_2 R_2 = C_1 R_1 \qquad (3.15)$$

3.5 Magnetic Forces

3.5.1 Magnetic fields

When a dc current flows in a conductor, a circular magnetic field is produced around the conductor as shown in Fig. 3.11a. Magnetic fields have magnetic flux or lines of force associated with them. When a current is passed through an

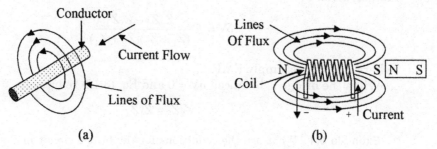

Figure 3.11 Magnetic lines of flux produced by (a) a straight conductor and (b) a coil.

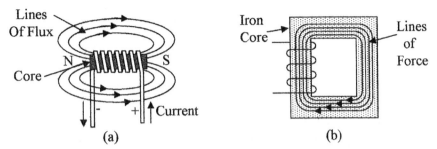

Figure 3.12 Shown are the lines of flux in (a) an iron cored coil and (b) in a closed iron core.

inductance or coil the magnetic field from each conductor adds to form a magnetic field as shown in Fig. 3.11b, similar to that of a magnet.

If a magnet is placed as shown with its north pole near the south pole of the coil, it will be attracted toward the coil. If the magnet is reversed, it will be repelled by the coil, i.e., opposite poles attract and like poles repel. When the current is turned off, the magnetic field in the coil starts to collapse and in doing so induces a voltage (reverse voltage) into the coil in the opposite direction to the initial driving voltage, to oppose the collapse of the current.

The ease of establishing the magnetic lines of force when a voltage is applied across an inductor is a measure of its inductance. Materials such as soft iron or ferrite are conductors of magnetic lines of force, so that the strength of the magnetic field and coil inductance are greatly increased if one of these materials is used as the core of the coil. The inductance will change as the core is moved in and out of the coil (see Fig. 3.12a). Hence, we have a means of measuring motion. Magnetic lines of force form a complete circuit as shown in Fig. 3.12a, the resistance to these lines of force (reluctance) can be reduced by providing a path for them both inside and outside the coil with a soft iron core as shown in Fig. 3.12b, with the reduced reluctance the magnet flux is increased several orders of magnitude in the closed magnetic core.

If the magnet in Fig. 3.11b is replaced by a second coil as shown in Fig. 3.13a and there is an ac current flowing through the first coil, then the build up, collapse,

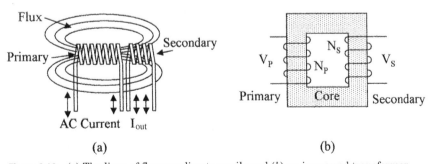

Figure 3.13 (a) The lines of flux coupling two coils and (b) an iron cored transformer.

and reversal of the magnetic line of flux in the first coil will also encompass and induce an ac electromotive force in the second coil.

Transformers are devices that transfer power from one coil (primary) to a second coil (secondary) as in Fig. 3.13*b*. The lines of flux are concentrated in the core, and because of the reduced reluctance with the complete core there is a very efficient and tight coupling between the two coils. This enables power transfer from one coil to another with low loss (>95 percent efficiency) without having any direct electrical connection between the two coils, as shown. This also allows for the transfer of power between different dc levels. By adjusting the ratio of turns between the two coils the voltage output from the secondary coil can be increased or reduced. The voltage relationship between the coils is given by

$$\frac{V_P}{V_S} = \frac{N_P}{N_S} \tag{3.16}$$

where V_P = primary voltage
N_P = number of turns on the primary coil
V_S = secondary voltage
N_S = number of turns on the secondary coil

Power and current can also be considered, in the case of power:

$$P_{in} = P_{out} \tag{3.17}$$

and in the case of current:

$$\frac{I_P}{I_S} = \frac{N_S}{N_P} \tag{3.18}$$

Example 3.5 A transformer with a primary of 1500 turns is used to generate 10 V ac from a supply voltage of 120 V. How many turns are there on the secondary? If the secondary is loaded with a 22 Ω resistor, what is the primary current?

$$N_S = \frac{1500 \times 10}{120} = 125 \text{ turns}$$

$$I_P = \frac{10 \times 125}{22 \times 1500} \text{ A} = 0.038 \text{ A} = 38 \text{ mA}$$

If instead of using an ac magnetic field to induce an EMF in a coil as in a transformer, a coil is rotated through a fixed magnetic field, an EMF will be induced in the coil. This is the basis for an ac generator, which consists of a rotating permanent magnet armature surrounded by fixed field coils. The lines of force from the magnetic field of the armature continually reverse in the field coils generating an ac EMF.

3.5.2 Analog meter

The attractive and repulsive forces between a permanent magnet field and the field produced by a dc-current-carrying coil are used in analog meter movements. If a current carrying coil that is free to rotate is placed in a magnetic

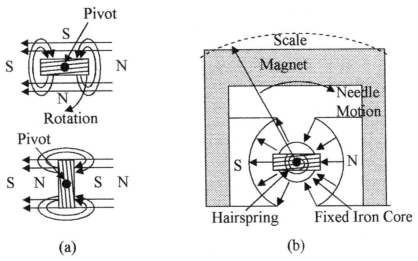

Figure 3.14 Shown in (a) are coils in a magnetic field and in (b) D'Arsonval meter movement.

field, as shown in Fig. 3.14a top diagram, the forces produced by the interaction between the magnetic fields will rotate the coil to the position shown at the bottom of the diagram, i.e., the magnetic fields will try to align themselves, so that the south pole of the coil is aligned to the north pole of the magnet and so on. This principle is used in the D'Arsonval meter movement shown in Fig. 3.14b. The poles of the permanent magnet are circularly shaped, with a fixed cylindrical soft iron core between them; this gives a very uniform radial magnetic field as shown. A coil is placed in the magnetic field as shown and is free to rotate about the soft iron core on low-friction bearings. Movement of the coil to align itself with the permanent magnetic field when a current is passed through it is opposed by hairsprings. The hairsprings are also used as electrical connections between the coil and the fixed electrical terminals.

The magnet field produced by the coil is directly proportional to the dc current flowing through the coil, and its deflection is therefore directly proportional to the current. A pointer attached to the coil gives the deflection of the coil on a scale. In such a meter full-scale deflection can be obtained with a current of about 50 µA. Current scales are obtained by using resistors to shunt some of the current around the meter movement, and voltages can be measured using series resistors to give high input impedances. AC voltages and currents are rectified, and then the dc is measured.

3.5.3 Electromechanical devices

Electromechanical devices use the magnetic forces developed in iron-cored coils for their operation. These forces can be very large when high currents are used in devices such as large motors. Electromechanical devices include

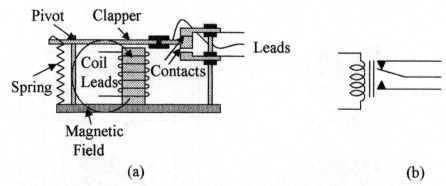

Figure 3.15 Cross section of (*a*) an electromagnetic relay and (*b*) a schematic diagram of the relay.

relays and actuators. Figure 3.15*a* shows an electromagnetic relay. When the coil is not energized the center lead is held in contact with the top lead by the spring as shown. When a current flows in the coil it is energized and a magnetic field is established. This attracts the clapper toward the coil and moves the contact arm downward breaking the contact between the upper and center leads and establishing a contact between the center and lower leads. The electromechanical relay in newer equipment is being replaced by solid state relays, which can use opto-isolation techniques when voltage isolation is required. Some of these devices and their use in process control will be discussed in more detail in later chapters. The schematic diagram of the relay is shown in Fig. 3.15*b*.

Summary

This chapter introduces the basic passive components in ac electricity. The main points discussed in this chapter are:

1. The effect of capacitance combined with resistance, and inductance combined with resistance on the phase relationship between current and voltage
2. The generation of time constants in ac circuits
3. Combination of resistance, capacitance, and inductance in a circuit, and its effect on phasing between current and voltage, calculation of circuit impedances, and the use of vectors for combining out-of-phase voltages and currents
4. Circuit impedances at resonance and their effects on circuit currents
5. The concept and use of filters for frequency selection and noise reduction
6. The ac Wheatstone bridge and its use for measuring the impedance of components
7. The generation of magnetic fields in straight conductors and coils and how the flux fields are enhanced by the reduction of resistance to the magnetic fields using iron cores

8. Tight magnetic coupling between coils and how this is used in transformers
9. Motion caused by attractive and repulsive magnetic fields and how this is used in meter movements

Problems

3.1 A voltage time constant of 15 ms is required. What value of resistance is needed if a capacitance of 0.1 μF is used?

3.2 A voltage step of 18 V is applied to the RC circuit shown in Fig. 3.1a. What will be the voltage across the capacitor after a time interval of $2CR$?

3.3 If an inductance of 21 mHz is used to obtain a current time constant of 12.5 μs, what value of resistance is required?

3.4 What is the time constant of a resistance of 18.5 kΩ in series with an inductance of 585 μHz?

3.5 In the series circuit shown in Fig. 3.16a the voltage measured across the inductance was 16.8 V, the voltage across the capacitor was 9.5 V. What is the voltage across the resistor?

3.6 What is the frequency of the ac source shown in Fig. 3.16a?

3.7 What is the value of the capacitor shown in Fig. 3.16a?

3.8 What is the resonant frequency of the series circuit in Fig. 3.16a?

3.9 What is the source current at resonance in Fig. 3.16a?

3.10 If the input ac frequency is 23.5 kHz, what is the current flowing in the circuit in Fig. 3.16a?

3.11 In the circuit shown in Fig. 3.16b, what is the current flowing in the resistor?

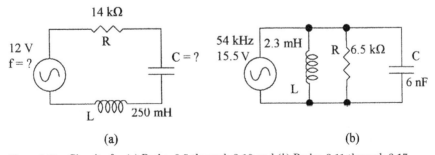

Figure 3.16 Circuits for (a) Probs. 3.5 through 3.10 and (b) Probs. 3.11 through 3.17.

3.12 In the circuit shown in Fig. 3.16b, what is the current flowing in the capacitor?

3.13 In the circuit shown in Fig. 3.16b, what is the current flowing in the inductance?

3.14 In Fig. 3.16b, how much current is supplied by the source?

3.15 What is the resonant frequency of the circuit in Fig. 3.16b?

3.16 In the circuit shown in Fig. 3.16b, what is the current flowing in the capacitor at resonance?

3.17 In the circuit shown in Fig. 3.16b, what is the current supplied by the source at resonance?

3.18 A 100 percent efficient iron-cored transformer has a primary winding with 1820 turns and a secondary winding with 125 turns. If the ac voltage across the primary winding is 110 V, what will be the ac voltage across the secondary?

3.19 The transformer in Prob. 3.18 has a resistive load of 2.2 kΩ across its secondary. What impedance does the transformer present to the source?

3.20 A transformer has 875 turns on the primary and 133 turns on the secondary. If the current flowing in the secondary is 3.45 A, what will be the current in the primary?

Chapter 4

Electronics

Chapter Objectives

This chapter will help you understand active devices, how they are used in instrument applications, and the difference between analog and digital circuits.

In this chapter the following points are discussed.

- The terms active and passive as applied to electronic components
- Signal amplification, gain adjustment, and feedback in amplifiers
- Operation of amplifiers
- Different types of amplifiers
- The difference between digital and analog circuits
- The instrument amplifier
- Introduction to digital circuits
- Conversion of analog signals to digital signals

The output from measuring or sensing devices is usually converted into electrical signals, so that they can be transmitted to a remote controller for processing and eventual actuator control or direct actuator control. Consequently, as well as understanding the operation of measuring and sensing devices, it is necessary to understand electricity and electronics as applied to signal amplification, control circuits, and the transmission of electrical signals. Measurable quantities are analog in nature, thus sensor signals are usually analog signals but can sometimes be converted directly into digital signals. Transmission of information over long distances can use analog or digital signals. Because of the higher integrity of digital signals compared to analog signals, and the fact that processors use digital signals, this form of transmission is preferred. Since both forms of transmission are in common use, an understanding of both signal forms is needed and will be discussed.

4.1 Introduction

Passive components—resistors, capacitors, and inductors—were studied in Chaps. 2 and 3. This chapter deals with active components, i.e., devices such as bipolar or metal oxide semiconductor (MOS) transistors, which are active devices and can amplify signals. Collectively all of these devices are referred to as electronics. Transistors are manufactured on semiconductor material (silicon or gallium arsenide) called chips or integrated circuits. Many tens of bipolar or thousands of MOS transistors can be made and interconnected on a single chip to form a complete complex circuit function or system.

4.2 Analog Circuits

The study of electronic circuits, where the inputs and outputs are continually varying, is known as *analog electronics*.

4.2.1 Discrete amplifiers

Transistors can be used to make discrete amplifiers. Figure 4.1 shows the circuit of a discrete bipolar (NPN type) and MOS (N channel) amplifier for amplifying ac signals. The difference in the levels of the dc input and output operating points, combined with the temperature drift, requires capacitive isolation between each stage, as well as with the application of direct resistive feedback. For instance, if the discrete devices were supplied from a 9 V supply, the dc input bias level would be about 3 V, and for symmetry of output the output dc level would be of about 6 V.

A bipolar device is a current amplifier, and its gain is given by

$$\text{Gain } (\beta) = \Delta I_C \text{ (collector } I \text{ change)}/\Delta I_B \text{ (base } I \text{ change)} \quad (4.1)$$

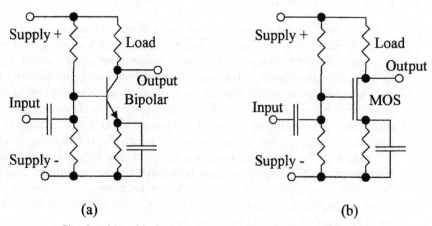

Figure 4.1 Circuits of (*a*) a bipolar discrete amplifier and (*b*) a MOS discrete amplifier.

A MOS device has a transconductance (transfer admittance) which is the change in the output current for a change in the input voltage. The transconductance is given by

$$\text{Transconductance } (\mu) = \Delta I_S \text{ (source } I \text{ change)}/\Delta E_G \text{ (gate } V \text{ change)} \quad (4.2)$$

Both β and μ, and other device parameters are temperature dependant, so that gain and operating point will vary with temperature.

Example 4.1 In Fig. 4.1b the MOS device has a transfer admittance of 4.5 mA/V. If the load resistance is 5 kΩ, what is the stage gain?

$$\text{Stage gain} = \mu \times 5 \text{ k}\Omega = 4.5 \text{ mA/V} \times 5 \text{ k}\Omega = 22.5$$

Sensor signals are normally low-level dc signals, which have to be amplified before they can be transmitted to a central control unit, or used to operate indicators or actuators. Discrete amplifiers are not suitable for sensor signal amplification because of their temperature drift and variations in stage gain.

4.2.2 Operational amplifiers

The integrated circuit made it possible to interconnect multiple active devices on a single chip to make an operational amplifier (op-amp), such as the LM741/107 general purpose op-amp. These amplifier circuits are small—one, two, or four can be encapsulated in a single plastic dual inline package (DIP) or similar package (see Fig. 4.2a). All of the discrete devices in an integrated circuit are manufactured as a group, giving all of them similar characteristics, and as they are in close proximity, they are at the same temperature. Thus, the integrated op-amp can be designed to overcome most of the problems encountered in the discrete device amplifier. This is achieved by using pairs of devices to balance each other's characteristics, minimizing temperature drift, and complementary pairs to reestablish dc operating levels. The end result is a general purpose amplifier that has high gain and low dc drift, so that it can amplify dc as well as ac signals. When the inputs are at 0 V, the output voltage is 0 V, or

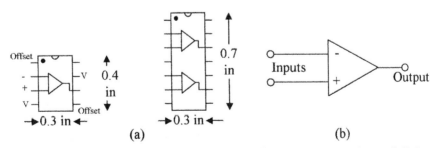

Figure 4.2 LM741/107 packages (a) 8-pin and 14-pin DIP showing connections and (b) Op-amp symbol.

can easily be adjusted to be 0 V with the offset null adjustment. Op-amps require a minimal number of external components. Direct feedback is easy to apply, giving stable gain characteristics and the output of one amplifier can be fed directly into the input of the next amplifier. Op-amps have dual inputs, one of which is a positive input, i.e., the output is in phase with the input; and the other is a negative input, i.e., the output is inverted from the input, so that depending on the input used, these devices can have an inverted or noninverted output and can amplify differential sensor signals or can be used to cancel electrical noise, which is often the requirement with low-level sensor signals. Op-amps are also available with dual outputs, i.e., both positive and negative going outputs are available. Op-amps are available in both bipolar and MOS technology.

Typical specifications for a general purpose integrated op-amp are:

Voltage gain	200,000
Output impedance	75 Ω
Input impedance bipolar	2 MΩ
Input impedance MOS	10^{12} Ω

The amplifier characteristics that enter into circuit design are

Input offset voltage	The voltage that must be applied between the inputs to drive the output voltage to zero.
Input offset current	The input current required to drive the output voltage to zero.
Input bias current	Average of the two input currents required to drive the output voltage to zero.
Slew rate	The rate of rise of the output voltage (V/μs) when a step voltage is applied to the input.
Unity gain frequency bandwidth	As the input frequency to an amplifier is increased the gain bandwidth decreases. This is the frequency where the voltage gain is unity.

The schematic representation of an op-amp is shown in Fig. 4.2b.

The specification and operating characteristics of bipolar operational amplifiers such as the LM 741/107 and MOS general purpose and high-performance op-amps can be found in semiconductor manufactures catalogs.

Many amplifiers use an offset control when amplifying small signals to set the dc output of the amplifier to zero when the dc input is zero. In the case of the LM 741/107 this is achieved by connecting a potentiometer (47 k) between the offset null points and taking the wiper to the negative supply line, as shown in Fig. 4.3.

In Fig. 4.4a the op-amp is configured as an inverting voltage amplifier. Resistors R_1 and R_2 provide feedback, i.e., some of the output signal is fed back to the input. The large amplification factor in op-amps tends to make some of them unstable and causes a dc drift of the operating point with temperature.

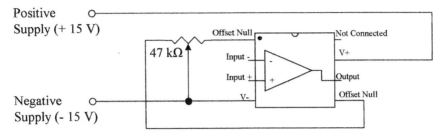

Figure 4.3 Offset control for the LM 741/107 op-amp.

The feedback stabilizes the amplifier, minimizes dc drift, and sets the gain to a known value.

When a voltage input signal is fed into the negative terminal of the op-amp, as in Fig. 4.4a, the output signal will be inverted. In this configuration for a high-gain amplifier, the voltage gain of the stage approximates to

$$\text{Gain} = \frac{-E_{out}}{E_{in}} = \frac{-R_2}{R_1} \quad (4.3)$$

The voltage gain of the amplifier can be adjusted with different values of R_2 or can be varied by adding a potentiometer in series with R_2. When the input signal is fed into the positive terminal the circuit is noninverting; such a configuration is shown in Fig. 4.4b. The voltage gain in this case approximates to

$$\text{Gain} = \frac{E_{out}}{E_{in}} = 1 + \frac{R_2}{R_1} \quad (4.4)$$

In this configuration the amplifier gain is 1 plus the resistor ratio, so that the gain does not vary directly with the resistor ratio. This configuration does, however, give a high-input impedance (that of the op-amp) and a low-output impedance.

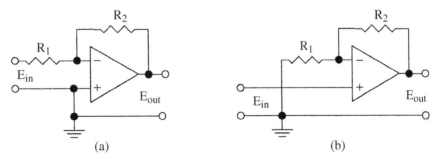

Figure 4.4 Circuit diagrams of (a) inverting amplifier and (b) noninverting amplifier.

Example 4.2 If in Fig. 4.4a, resistor $R_1 = 1200\ \Omega$ and resistor $R_2 = 150$ kΩ, what is the gain, and what is the output voltage amplitude if the ac input voltage is 3.5 mV?

$$\text{Gain} = \frac{R_2}{R_1} = \frac{150}{1.2} = 125$$

ac output voltage $= -3.5 \times 125$ mV $= -437.5$ mV $= -0.44$ V

Example 4.3 In Fig. 4.4a and b, $R_1 = 4.7$ kΩ and $R_2 = 120$ kΩ. If a dc voltage of 0.15 V is applied to the inputs of each amplifier, what will be the output voltages?

In Fig. 4.4a

$$V_{out} = \frac{-120 \times 0.15\ \text{V}}{4.7} = -3.83\ \text{V}$$

In Fig. 4.4b

$$V_{out} = \left(1 + \frac{120}{4.7}\right) 0.15\ \text{V} = +3.98\ \text{V}$$

The circuits shown in Fig. 4.4 were for voltage amplifiers. Op-amps can also be used as current amplifiers, voltage to current and current to voltage converters, and special-purpose amplifiers. In Fig. 4.5a the op-amp is used as a current-to-voltage converter. When used as a converter, the relation between input and output is called the transfer function μ (or ratio). These devices do not have gain as such because of the different input and output units. In Fig. 4.5a the transfer ratio is given by

$$\mu = \frac{-E_{out}}{I_{in}} = R_1 \qquad (4.5)$$

Example 4.4 In Fig. 4.5a the input current is 165 µA and the output voltage is -2.9 V. What is the transfer ratio and the value of R_1?

$$\mu = \frac{2.9\ \text{V}}{165\ \mu\text{A}} = 17.6\ \text{V/mA} = 17.6\ \text{kV/A}$$

$$R_1 = \frac{2.9\ \text{V}}{165\ \mu\text{A}} = 17.6\ \text{k}\Omega$$

In Fig. 4.5b the op-amp is used as a voltage-to-current converter. In this case the transfer ratio is given by

$$\frac{I_{out}}{E_{in}} = \frac{-R_2}{R_1 R_3}\ \text{mhos} \qquad (4.6)$$

Note, in this case the units are in mhos (1/ohms), and the resistors are related by the equation

$$R_1(R_3 + R_5) = R_2 R_4 \qquad (4.7)$$

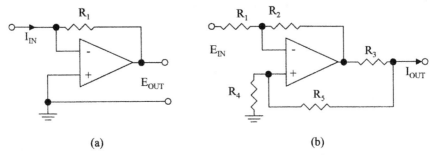

Figure 4.5 Examples of (a) a current to voltage converter and (b) a voltage to current converter.

Example 4.5 In Fig. 4.5b, $R_1 = R_4 = 5$ kΩ and $R_2 = 100$ kΩ. What is the value of R_3 and R_5 if the op-amp is needed to convert an input voltage of 3 V to an output of 20 mA?

$$\frac{I_{out}}{E_{in}} = \frac{R_2}{R_1 R_3} = \frac{20 \times 10^{-3}}{3} = 6.67 \times 10^{-3}$$

$$\frac{100 \times 10^3}{5 \times 10^3 \times R_3} = 6.67 \times 10^{-3}$$

$$R_3 = \frac{20}{6.67 \times 10^{-3}} = 3 \text{ k}\Omega$$

$$R_5 = \frac{R_2 R_4}{R_1 - R_3} = R_2 - R_3 = (100 - 3) \text{ k}\Omega = 97 \text{ k}\Omega$$

The above circuit configurations can be used for dc as well as ac amplification or conversion, the only difference being that capacitors are normally used in ac amplifiers between stages to prevent any dc offset levels present from affecting the biasing or operating levels of the following op-amps. The dc operating point of an ac amplifier can be set at the input to the stage. High-gain dc amplifiers are directly coupled and use special op-amps that have a low drift with temperature changes. The voltage supplies to the op-amp are regulated to minimize output changes with supply voltage variations.

4.2.3 Current amplifiers

Devices that amplify currents are referred to as current amplifiers. However, in industrial instrumentation a voltage-to-current converter is sometimes referred to as a current amplifier. Figure 4.6a shows a basic current amplifier. The gain is given by

$$\frac{I_{out}}{I_{in}} = \frac{R_2 R_6}{R_1 R_3} \qquad (4.8)$$

where the resistors are related by the equation

$$R_1 (R_3 + R_5) = R_2 R_4 \qquad (4.9)$$

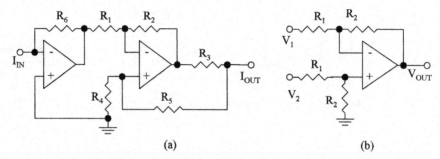

Figure 4.6 Circuit diagram of the basic configuration of (a) a current amplifier and (b) a differential amplifier.

4.2.4 Differential amplifiers

A differential amplifier is a dual input amplifier that amplifies the difference between two signals, such that the output is the gain multiplied by the magnitude of the difference between the two signals. One signal is fed to the negative input of the op-amp and the other signal is fed to the positive input of the op-amp. Hence the signals are subtracted before being amplified. Figure 4.6b shows a basic differential voltage amplifier. The output voltage is given by

$$V_{out} = \frac{R_2}{R_1}(V_2 - V_1) \qquad (4.10)$$

Signals can also be subtracted or added in a resistor network prior to amplification.

Example 4.6 In the dc amplifier shown in Fig. 4.7a, an input of 130 mV is applied to terminal A, and -85 mV is applied to terminal B. What is the output voltage (assume the amplifier was zeroed with 0 V at the inputs)?

$$E_{out} = \frac{-\Delta V_{in} \times 120}{4.7} = [-130 + (-85)] \text{ mV} \times \frac{120}{4.7} = -0.215 \times 25.5 \text{ V} = -5.5 \text{ V}$$

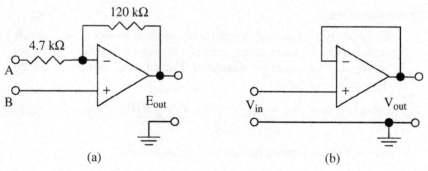

Figure 4.7 Schematic diagrams for (a) Example 4.6 and (b) a buffer amplifier.

4.2.5 Buffer amplifiers

An impedance matching op-amp is called a buffer amplifier. Such amplifiers have feedback to give unity voltage gain, high input impedance (many megaohms), and low output impedance (<20 Ω), such an amplifier is shown in Fig. 4.7b. In this context impedance is used to cover both ac impedance and dc resistance. Circuits have both input and output impedance.

The effect of loading on a circuit can be seen in Fig. 4.8a. The resistor divider gives an output voltage of 8 V and an output impedance of 2.7 kΩ (effectively this impedance is 4 kΩ in parallel with 8 kΩ). If this divider is loaded with a circuit with an input impedance of 2 kΩ, the output voltage will drop from 8 to 3.43 V. A buffer amplifier can be used as shown in Fig. 4.8b to match the input impedance of the second circuit to the first circuit, thus giving an output voltage of 8 V across the 2 kΩ load.

Example 4.7 In Fig. 4.8b, what is the output voltage of the buffer amplifier? Assume the input impedance of the buffer amplifier is 2 MΩ and its output impedance is 15 Ω. 2 MΩ in parallel with 8 kΩ has an effective resistance of 7.97 kΩ

$$\text{Voltage at input to buffer} = \frac{12 \times 7.97}{7.97 + 4} \text{ V} = 7.99 \text{ V}$$

From this we get that the buffer loading reduces the output voltage from the resistive divider by 0.01 V, which is about 0.125 percent. The output impedance of the buffer is effectively in series with the 2-kΩ load, so that the output voltage E_{out} is given by

$$E_{out} = \frac{7.99 \times 2000}{2000 + 15} \text{ V} = 7.93 \text{ V}$$

Thus, the total loading effect is a reduction of 0.07 V in 8 V, or about 0.9 percent compared to 57.5 percent with direct loading. This error could be totally corrected if the amplifier had a gain of 1.01.

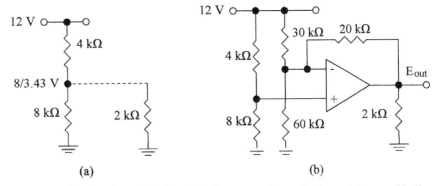

Figure 4.8 Circuits shows (a) effect of loading on a voltage divider and (b) use of buffer in Example 4.7.

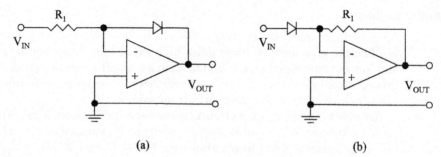

Figure 4.9 Circuits of nonlinear amplifiers: (a) log amplifier and (b) antilog amplifier.

4.2.6 Nonlinear amplifiers

Many sensors have a logarithmic or nonlinear transfer characteristic and such devices require signal linearization. This can be implemented by using amplifiers with nonlinear characteristics. These are achieved by the use of nonlinear elements such as diodes or transistors in the feedback loop. Figure 4.9 shows two examples of nonlinear amplifiers using a diode in the feedback loop. In (a) the amplifier is configured as a logarithmic amplifier and in (b) the amplifier is configured as an antilogarithmic amplifier. Combinations of resistors and nonlinear elements can be chosen to match the characteristics of many sensors for linearization of the output from the sensor.

4.2.7 Instrument amplifier

Because of the very high accuracy requirement in instrumentation, the op-amp circuits shown in Fig. 4.4 are not ideally suited for low-level instrument signal amplification. The op-amp can have different input impedances at the two inputs, the input impedances can be relatively low and tend to load the sensor output, can have different gains at the inverting and noninverting inputs, and common mode noise can be a problem. Op-amps configured for use as an instrument amplifier is shown in Fig. 4.10. This amplifier has balanced

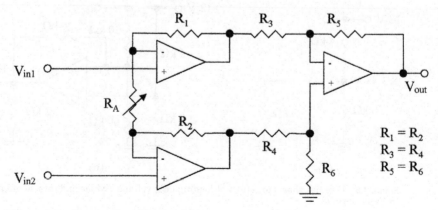

Figure 4.10 Circuit schematic of an instrumentation amplifier.

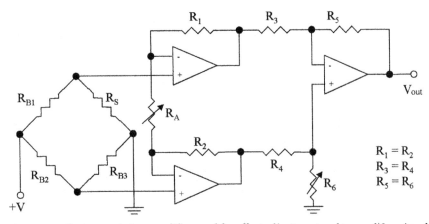

Figure 4.11 Instrumentation amplifier used for offset adjustment and to amplify a signal from a bridge.

inputs with very high input impedance and good common-mode noise reduction. Gain is set by R_A.

The output voltage is given by

$$V_{out} = \frac{R_5}{R_3}\left(\frac{2R_1}{R_A}+1\right)(V_{IN2}-V_{IN1}) \qquad (4.11)$$

Figure 4.11 shows a practical circuit using an instrumentation amplifier to amplify the output signal from a resistive bridge; R_6 is used to adjust for any zero signal offset.

4.2.8 Amplifier applications

In process control, amplifiers are used in many applications other than signal amplification, filtering, and linearization. Some of these applications are as follows:

Capacitance multiplier

Gyrator

Sine wave oscillators

Power supply regulators

Level detection

Sample and hold

Voltage reference

Current mirrors

Voltage-to-frequency converters

Voltage-to-digital converters

Pulse amplitude modulation

More information on the design and use of these circuits can be found in analog electronic text books and the like.

4.3 Digital Circuits

The study of electronic circuits where the inputs and outputs are limited to two fixed or discrete values or logic levels is called digital electronics. Digital technology would take many volumes to do it justice, so in this text we can only scratch the surface. There is a place for both analog and digital circuits in instrumentation. Sensors and instrumentation functions are analog in nature. However, the digital circuits have many advantages over analog circuits. Analog signals are easily converted to digital signals using commercially available analog-to-digital converters (ADC). In new designs, digital circuits will be used wherever possible.

Some of the advantages of digital circuits are

- Lower power requirements
- More cost effective
- Can transmit signals over long distances without loss of accuracy and elimination of noise
- High-speed signal transmission
- Memory capability for data storage
- Controller and alpha numeric display compatible

4.3.1 Digital signals

Digital signals are either high or low logic levels. Most digital circuits use a 5-V supply. The logic low (binary 0) level is from 0 to 1V, the logic high (binary1) level is from 2 to 5V; 1 to 2 V is an undefined region, i.e., any voltage below 1 V is considered a 0 level and any voltage above 2 V is considered a 1 level. In circuits where the supply voltage is other than 5 V, a 0 level is still considered as a 0 V level or the output drivers are sinking current, i.e., connecting the output terminal to ground, and a 1 level is close to the supply voltage or the output drivers are sourcing current, i.e., connecting the output terminal to the supply rail.

4.3.2 Binary numbers

We use the decimal system (base 10) for mathematical functions, whereas electronics uses the binary system (base 2) to perform the same functions. The rules are the same when performing calculations using either of the two numbering systems (to the base 10 or 2). Table 4.1 gives a comparison between counting in the decimal and binary systems. The least significant bit (LSB) or unit number is the right-hand bit. In the decimal system when the unit numbers are used we go to the tens, that is, 9 goes to10, and when the tens are used we go to the hundreds, that is, 99 goes to 100 and so forth. The binary system is the same when the 0 and 1 are used in the LSB position, then we go to the next position and so on, that is, 1 goes to 10,11 goes to 100, and 111 goes to 1000, and

TABLE 4.1 Decimal and Binary Equivalents

Decimal	Binary	Decimal	Binary
0000	0000000	0021	0010101
0001	0000001	—	—
0002	0000010	0031	0011111
0003	0000011	0032	0100000
0004	0000100	—	—
0005	0000101	0063	0111111
—	—	0064	1000000
0007	0000111	—	—
0008	0001000	0099	1100011
0009	0001001	0100	1100100
0010	0001010	0101	1100101
0011	0001011	0999	1111100111
0015	0001111	1000	1111101000
0016	0010000	1001	1111101001
—	—	1002	1111101010
0020	0010100	1024	10000000000

so forth. The only difference is that, to represent a number it requires more digits when using a binary system than in the decimal system.

Binary numbers can be easily converted to decimal numbers by using the power value of the binary number. Table 4.2 gives the power value of binary numbers versus their location from the LSB and their decimal equivalent.

Note that when counting locations, the count starts at 0 and not, as might be expected, at 1.

Each binary digit is called a bit, 4 bits are defined as a nibble, 8 bits form a byte, and 2 bytes or 16 bits are called a word. A word is often broken down into 4 nibbles, where each nibble is represented by a decade number plus letters as shown in Table 4.3. Thus, a word can be represented by 4 decade numbers plus the first six letters of the alphabet. This system is known as the hexadecimal system.

TABLE 4.2 Power Value of Binary Numbers

Location	8	7	6	5	4	3	2	1	0
Power value	2^8	2^7	2^6	2^5	2^4	2^3	2^2	2^1	2^0
Decimal number	256	128	64	32	16	8	4	2	1

Example 4.8 What is the decimal number equivalent of the binary number 101100101? The equivalent power values are given by

Binary number	1		0	1	1	0	0	1	0	1
Location	8			6	5			2		0
Power value	2^8			2^6	2^5			2^2		2^0
Decimal number	256	+		64	+ 32		+	4	+	1
Decimal Total	=		357							

TABLE 4.3 Numbering Equivalent in the Hexadecimal (H) System

Binary number	Decade equivalent	Binary number	Decade equivalent
0000	0	1000	8
0001	1	1001	9
0010	2	1010	A
0011	3	1011	B
0100	4	1100	C
0101	5	1101	D
0110	6	1110	E
0111	7	1111	F

Example 4.9 What is the hexadecimal equivalent of the binary word 1101001110110111?
The binary word is broken down into groups of 4 bits (byte) starting from the LSB and going to the most significant bit (MSB).

	MSB			LSB
Word separated into bytes	1101	0011	1011	0111
Hexadecimal equivalent	D	3	B	7
Decimal number equivalent	54,199			

Binary circuits are synchronized by clock signals which are referenced to very accurate crystal oscillators ($< \pm 0.001$ percent), using counters and dividers. The clock signal can be used to generate very accurate delays and timing signals, compared to RC-generated delays and timing which can have tolerances of $> \pm 10$ percent, so that delays and timing will be done almost entirely by digital circuits in new equipment.

4.3.3 Logic circuits

The basic building blocks in digital circuits are called gates. These are buffer, inverter, AND, NAND, OR, NOR, XOR, and XNOR. These basic blocks are interconnected to build functional blocks such as encoders, decoders, adders, counters, registers, multiplexers, demultiplexers, memories, and the like. The functional blocks are then interconnected to make systems, i.e., calculators, computers, microprocessors, clocks, function generators, transmitters, receivers, digital instruments, ADC and digital-to-analog converters (DAC), telephone systems and the like, to name a few.

Figure 4.12a shows the circuit of a complementary MOS (CMOS) inverter. The circuit uses both N- and P-channel complementary devices (note device symbols). Figure 4.12b shows the equivalent gate symbol. When the input to the gate is low (0) the positive-channel MOS (PMOS) is "ON" and the negative MOS (NMOS) is "OFF" so that the output is held high (1), and when the input is high (1) the PMOS is "OFF" and the NMOS is "ON", which will hold the output low (0), so that the input sign is inverted at the output. One of the MOS devices is always "OFF", so that the circuit draws no current from the supply (except during switching) making CMOS circuits very power efficient.

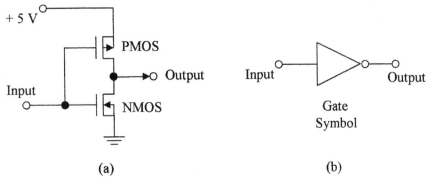

Figure 4.12 Circuit components used to make (a) a MOS inverter and (b) an inverter symbol.

4.3.4 Analog-to-digital conversion

The amplitude of an analog signal can be represented by a digital number, for instance, an 8-bit word can represent numbers up to 255, so that it can represent an analog voltage or current with an accuracy of 1 in 255 (assuming the conversion is accurate to 1 bit) or 0.4 percent accuracy. Similarly a 10 and 12-bit word can represent analog signals to accuracies of 0.1 and 0.025 percent, respectively.

Commercial integrated A/D converters are readily available for instrumentation applications. Several techniques are used for the conversion of analog signals–to digital signals. These are

Flash converters which are very fast and expensive with limited accuracy, that is, 6-bit output with a conversion time of 33 ns. The device can sample an analog voltage 30 million times per second.

Successive approximation is a high-speed, medium-cost technique with good accuracy, that is, the most expensive device can convert an analog voltage to 12 bits in 20 µs, and a less expensive device can convert an analog signal to 8 bits in 30 µs.

Resistor ladder networks are used in low-speed, medium-cost converters. They have a 12-bit conversion time of about 5 ms.

Dual slope converters are low-cost, low-speed devices but have good accuracy and are very tolerant of high noise levels in the analog signal. A 12-bit conversion takes about 20 ms.

Analog signals are constantly changing, so that for a converter to make a measurement, a sample-and-hold technique is used to capture the voltage level at a specific instant in time. Such a circuit is shown in Fig. 4.13a, with the waveforms shown in Fig. 4.13b. The N-channel field effect transistor (FET) in the sample-and-hold circuit has a low impedance when turned "ON" and a very high impedance when turned "OFF". The voltage across capacitor C follows the input analog voltage when the FET is "ON" and holds the dc level of the analog voltage when the FET is turned "OFF". During the "OFF" period the ADC measures the dc level of the analog voltage and converts it into a digital signal. As the

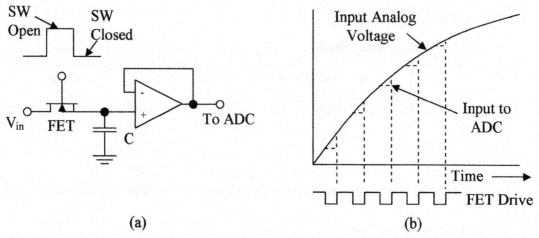

Figure 4.13 (a) Sample and hold circuit and (b) waveforms for the circuit.

sampling frequency of the ADC is much higher than the frequency of the analog signal, the varying amplitude of the analog signal can be represented in a digital format during each sample period and stored in memory. The analog signal can be regenerated from the digital signal using a DAC.

Figure 4.14a shows the block diagram of the ADC 0804, a commercial 8-bit ADC. The analog input is converted to a byte of digital information every few milliseconds.

An alternative to the ADC is the voltage-to-frequency converter. In this case the analog voltage is converted to a frequency. Commercial units such as the LM 331 shown in Fig. 4.14b are available for this conversion. These devices

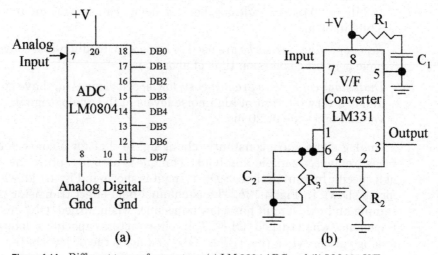

Figure 4.14 Different types of converters: (a) LM 0804 ADC and (b) LM 331 V/F converter.

have a linear relation between voltage and frequency. The operating characteristics of the devices are given in the manufacturers' data sheets.

4.4 Circuit Considerations

Analog circuits can use either bipolar or MOS-integrated amplifiers. Bipolar op-amps tend to have a lower input dc offset, but MOS op-amps have higher input impedance. Since a large number of op-amps and special amplifiers are available, the manufacturers' data sheets should be consulted to decide which amplifier is best suited for a specific application.

Digital circuits can be divided by the number of components integrated onto a single silicon chip into small-scale, medium-scale, and large-scale integration. In small-scale integration (up to 100 devices) such as the SN 54/74 family of digital circuits, both bipolar and CMOS devices are used. This family of devices contains gates and small building blocks. Medium scale integration (over 100,000 devices) will use only CMOS devices. This is because of the excessive power requirements, high dissipation, and relatively large size of bipolar devices. Such circuits contain large building blocks on a single chip. Large-scale integrated circuits contain over 1,000,000 devices, and use CMOS technology. These devices are used for large memories, microprocessors, and microcontrollers, and such circuits can contain several million devices.

Systems containing a large number of gates are now using programmable logic arrays (PLA) to replace the SN 54/74 gate family of devices, as one of these devices can replace many gate devices, requires less power, and can be configured (programmed) by the end user to perform all the required system functions.

Summary

In this chapter, op-amps, their use in amplifying analog signals, and their use as signal converters were discussed. The relation between analog signals and digital signals, and the conversion of analog-to-digital signals was covered.

Salient points discussed in this chapter are:

1. Discrete amplifiers, their use in ac signal amplification, and why they are not suitable for dc signal amplification
2. The op-amp and its basic characteristics, its versatility and use in signal amplification, and methods of setting the zero operating point
3. Signal inversion and noninversion, methods of applying feedback for gain control and stability
4. Use of the op-amp as a signal converter, impedance matching, set zero control, and span adjustment
5. Configuration of op-amps to make an instrument amplifier for accurate signal amplification and noise reduction
6. Introduction to digital circuits plus a comparison between analog and digital circuits

7. Binary, hexadecimal, decimal equivalents, and conversions between the counting schemes
8. Logic circuits used in digital systems and circuit considerations
9. Conversion of analog signals to digital signals and the resolution obtained

Problems

4.1 What is the difference between analog and digital circuits?

4.2 What is the stage gain of the discrete amplifier shown in Fig. 4.15a if the MOS device has an admittance of 5.8 mA/V and the load resistance is 8.2 kΩ? Assume the capacitors have zero impedance.

4.3 What load must be used in the discrete amplifier shown in Fig. 4.15a if a stage gain of 33 is required? Assume the capacitors have zero impedance.

4.4 What is the stage gain of the op-amp shown in Fig. 4.15b if $R = 285$ kΩ?

4.5 What is the value of the feedback resistor needed for the op-amp shown in Fig. 4.14b to obtain a voltage gain of 533?

4.6 Redraw the inverting amplifier in Fig. 4.15b as a noninverting amplifier with a voltage gain of 470.

4.7 What is the offset null in an op-amp used for?

4.8 If the amplitude of the input in the amplifier shown in Fig. 4.15c is 14 µA. What is the amplitude of the output if the feedback resistor $R = 56$ kΩ?

4.9 What is the transfer ratio of the amplifier shown in Fig. 4.15c, if the feedback resistor $R = 27$ kΩ?

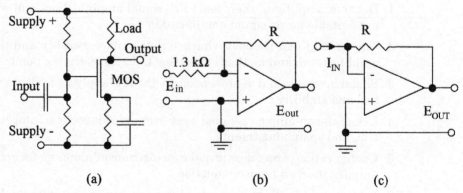

Figure 4.15 Circuits for use in (a) Probs. 4.2 and 4.3, (b) Probs. 4.4, 4.5, and 4.6, and (c) Probs. 4.8 and 4.9.

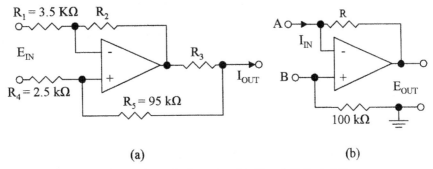

Figure 4.16 Circuits for use in (a) Prob. 4.10 and 4.11 and (b) Prob.4.12.

4.10 What is the output current if the input voltage is 3.8 mV in Fig. 4.16a? Assume $R_3 = 1.5$ kΩ.

4.11 What is the value of R_3 in Fig. 4.16a for a transfer ratio of 8.5 mA/μV? Assume $R_2 = 100$ kΩ

4.12 If in Fig. 4.16b, input A is 17 mV, and input B is −21 mV, what is the value of the output voltage if $R = 83$ kΩ?

4.13 What is the base number used in a binary circuit?

4.14 What is the decimal equivalent of 1011001?

4.15 What is the binary equivalent of 0037?

4.16 What is the hexadecimal equivalent of 011010011100?

4.17 What is the hexadecimal equivalent of 111000111010?

4.18 How many bits are there in a (a) byte and (b) word?

4.19 Does a "1" level represent an output that is sourcing or sinking current?

4.20 Name the types of gates used in digital circuits.

Chapter 5

Pressure

Chapter Objectives

This chapter will help you understand the units used in pressure measurements and become familiar with the most common methods of using the various pressure measurement standards.

Discussed in this chapter are

- The terms - pressure, specific weight, specific gravity (SG), and buoyancy
- The difference between atmospheric, absolute, gauge, and differential pressure values
- Various pressure units in use, i.e., British units versus SI (metric) units
- Various types of pressure measuring devices
- Difference in static, dynamic, and impact pressures
- Laws applied to pressure
- Application considerations

5.1 Introduction

Pressure is the force exerted by gases and liquids due to their weight, such as the pressure of the atmosphere on the surface of the earth and the pressure containerized liquids exert on the bottom and walls of a container.

Pressure units are a measure of the force acting over a specified area. It is most commonly expressed in pounds per square inch (psi), sometimes pounds per square foot (psf) in English units, or pascals (Pa or kPa) in metric units.

$$\text{Pressure} = \frac{\text{force}}{\text{area}} \qquad (5.1)$$

TABLE 5.1 Specific Weights and Specific Gravities of Some Common Materials

	Temperature, °F	Specific weight		Specific gravity
		lb/ft^3	kN/m^3	
Acetone	60	49.4	7.74	0.79
Alcohol (ethyl)	68	49.4	7.74	0.79
Glycerin	32	78.6	12.4	1.26
Mercury	60	846.3	133	13.55
Steel		490	76.93	7.85
Water	39.2	62.43	9.8	1.0

Conversion factors. 1 ft^3 = 0.028 m^3, 1 lb = 4.448 N, and 1 lb/ft^3 = 0.157 kN/m^3.

Example 5.1 The liquid in a container has a total weight of 250 lb; the container has a 3.0 ft^2 base. What is the pressure in pounds per square inch?

$$\text{Pressure} = \frac{250}{3 \times 144} \text{ psi} = 0.58 \text{ psi}$$

5.2 Basic Terms

Density ρ is defined as the mass per unit volume of a material, i.e., pound (slug) per cubic foot (lb (slug)/ft^3) or kilogram per cubic meter (kg/m^3).

Specific weight γ is defined as the weight per unit volume of a material, i.e., pound per cubic foot (lb/ft^3) or newton per cubic meter (N/m^3).

Specific gravity of a liquid or solid is a dimensionless value since it is a ratio of two measurements in the same unit. It is defined as the density of a material divided by the density of water or it can be defined as the specific weight of the material divided by the specific weight of water at a specified temperature. The specific weights and specific gravities of some common materials are given in Table 5.1. The specific gravity of a gas is its density/specific weight divided by the density/specific weight of air at 60°F and 1 atmospheric pressure (14.7 psia). In the SI system the density in g/cm^3 or Mg/m^3 and SG have the same value.

Static pressure is the pressure of fluids or gases that are stationary or not in motion (see Fig. 5.1). Point A is considered as static pressure although the fluid above it is flowing.

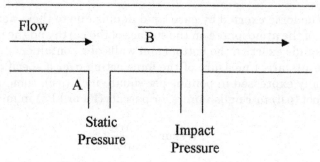

Figure 5.1 Illustration of static, dynamic, and impact pressures.

Dynamic pressure is the pressure exerted by a fluid or gas when it impacts on a surface or an object due to its motion or flow. In Fig. 5.1 the dynamic pressure is $(B - A)$.

Impact pressure (total pressure) is the sum of the static and dynamic pressures on a surface or object. Point B in Fig. 5.1 depicts the impact pressure.

5.3 Pressure Measurement

There are six terms applied to pressure measurements. They are as follows:

Total vacuum—which is zero pressure or lack of pressure, as would be experienced in outer space.

Vacuum is a pressure measurement made between total vacuum and normal atmospheric pressure (14.7 psi).

Atmospheric pressure is the pressure on the earth's surface due to the weight of the gases in the earth's atmosphere and is normally expressed at sea level as 14.7 psi or 101.36 kPa. It is however, dependant on atmospheric conditions. The pressure decreases above sea level and at an elevation of 5000 ft drops to about 12.2 psi (84.122 kPa).

Absolute pressure is the pressure measured with respect to a vacuum and is expressed in pounds per square inch absolute (psia).

Gauge pressure is the pressure measured with respect to atmospheric pressure and is normally expressed in pounds per square inch gauge (psig). Figure 5.2a shows graphically the relation between atmospheric, gauge, and absolute pressures.

Differential pressure is the pressure measured with respect to another pressure and is expressed as the difference between the two values. This would represent two points in a pressure or flow system and is referred to as the *delta p*

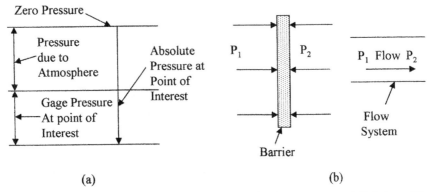

Figure 5.2 Illustration of (*a*) gauge pressure versus absolute pressure and (*b*) delta or differential pressure.

TABLE 5.2 Pressure Conversions

	Water		Mercury[‡]			
	in[*]	cm[†]	Mm	Ins	kPa	Psi
1 psi	27.7	70.3	51.7	2.04	6.895	1
1 psf	0.19	0.488	0.359	0.014	0.048	0.007
1 kPa	4.015	10.2	7.5	0.295	1	0.145
Atmospheres	407.2	1034	761	29.96	101.3	14.7
Torr	0.535	1.36	1	0.04	0.133	0.019
Millibar	0.401	1.02	0.75	0.029	0.1	0.014

[*]At 39°F.
[†]At 4°C.
[‡]Mercury at 0°C.

or Δp. Figure 5.2b shows two situations, where differential pressure exists across a barrier and between two points in a flow system.

Example 5.2 The atmospheric pressure is 14.5 psi. If a pressure gauge reads 1200 psf, what is the absolute pressure?

$$\text{Absolute pressure} = 14.5 \text{ psi} + \frac{1200 \text{ psf}}{144} = 14.5 \text{ psi} + 8.3 \text{ psi} = 22.8 \text{ psia}$$

A number of *measurement units* are used for pressure. They are as follows:

1. Pounds per square foot (psf) or pounds per square inch (psi)
2. Atmospheres (atm)
3. Pascals (N/m^2) or kilopascal (1000Pa)*
4. Torr = 1 mm mercury
5. Bar (1.013 atm) = 100 kPa

Table 5.2 gives a table of conversions between various pressure measurement units.

Example 5.3 What pressure in pascals corresponds to 15 psi?

$$p = 15 \text{ psi } (6.895 \text{ kPa/psi}) = 102.9 \text{ kPa}$$

5.4 Pressure Formulas

Hydrostatic pressure is the pressure in a liquid. The pressure increases as the depth in a liquid increases. This increase is due to the weight of the fluid above the measurement point. The pressure is given by

$$p = \gamma h \qquad (5.2)$$

Note: 1 N = force to accelerate 1 kg by 1 m/s^2 (units kg m/s^2), that is, 1 Pa = 1 N/m^2 = 1 kg m/s^2 ÷ g = 1 kg m/s^2 ÷ 9.8 m/s^2 = 0.102 kg/m^2, 1 dyn = 10^{-5} N, where g (gravitational constant) = 9.8 m/s^2 or 32.2 ft/s^2 and force = mass x acceleration.

where p = pressure in pounds per unit area or pascals
γ = the specific weight (lb/ft^3 in English units or N/m^3 in SI units)
h = distance from the surface in compatible units (ft, in, cm, m, and so on)

Example 5.4 What is the gauge pressure in (a) kilopascals and (b) newtons per square centimeter at a distance 1 m below the surface in water?

(a) $p = 100 \text{ cm/m}/10.2 \text{ cm/kPa} = 9.8 \text{ kPa}$
(b) $p = 9.8 \text{ N/m}^2 = 9.8/10{,}000 \text{ N/cm}^2 = 0.98 \times 10^{-3} \text{ N/cm}^2$

The pressure in this case is the gauge pressure, i.e., kPa(g). To get the total pressure, the pressure of the atmosphere must be taken into account. The total pressure (absolute) in this case is 9.8 + 101.3 = 111.1 kPa(a).

The g and a should be used in all cases to avoid confusion. In the case of pounds per square inch and pounds per square foot this would become pounds per square inch gauge and pounds per square foot gauge, or pounds per square inch absolute and pounds per square foot absolute. Also it should be noted that if glycerin was used instead of water the pressure would have been 1.26 times higher, as its specific gravity is 1.26.

Example 5.5 What is the specific gravity of mercury if the specific weight of mercury is 846.3 lb/ft^3?

$$SG = 846.3/62.4 = 13.56$$

Head is sometimes used as a measure of pressure. It is the pressure in terms of a column of a particular fluid, i.e., a head of 1 ft of water is the pressure that would be exerted by a 1 ft tall column of water, that is, 62.4 psfg, or the pressure exerted by 1 ft head of glycerin would be 78.6 psfg.

Example 5.6 What is the pressure at the base of a water tower which has 50 ft of head?

$$p = 62.4 \text{ lb/ft}^3 \times 50 \text{ ft} = 3120 \text{ psfg} = 3120 \text{ psf}/144 \text{ ft}^2/\text{in}^2 = 21.67 \text{ psig}$$

The hydrostatic paradox states that the pressure at a given depth in a liquid is independent of the shape of the container or the volume of liquid contained. The pressure value is a result of the depth and density. Figure 5.3a shows various shapes of tanks. The total pressure or forces on the sides of the container depend on its shape, but at a specified depth. The pressure is given by Eq. (5.2).

Buoyancy is the upward force exerted on an object immersed or floating in a liquid. The weight is less than it is in air due to the weight of the displaced fluid. The upward force on the object causing the weight loss is called the buoyant force and is given by

$$B = \gamma V \tag{5.3}$$

where B = buoyant force (lb)
γ = specific weight (lb/ft^3)
V = volume of the liquid displaced (ft^3)

If working in SI units, B is in newtons, γ in newton per cubic meter, and V in cubic meters.

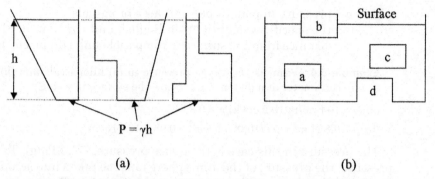

Figure 5.3 Diagrams demonstrating (a) the hydrostatic paradox and (b) buoyancy.

In Fig. 5.3b, a, b, c, and d are of the same size. The buoyancy forces on a and c are the same although their depth is different. There is no buoyancy force on d as the liquid cannot get under it to produce the force. The buoyancy force on b is half of that on a and c, as only half of the object is submersed.

Example 5.7 What is the buoyant force on a wooden cube with 3-ft sides floating in water, if the block is half submerged?

$$B = 62.4 \text{ lb/ft}^3 \times 3 \text{ ft} \times 3 \text{ ft} \times 1.5 \text{ ft} = 842.4 \text{ lb}$$

Example 5.8 What is the apparent weight of a 3-m^3 block of steel totally immersed in glycerin?

Weight of steel in air = 3×76.93 kN = 230.8 kN
Buoyancy force on steel = 3×12.4 kN = 37.2 kN
Apparent weight = $230.8 - 37.2 = 193.6$ kN (19.75 Mg)

Pascal's law states that the pressure applied to an enclosed liquid (or gas) is transmitted to all parts of the fluid and to the walls of the container. This is demonstrated in the hydraulic press in Fig. 5.4. A force F_S, exerted on the small piston (ignoring friction), will exert a pressure in the fluid which is given by

$$p = \frac{F_S}{A_S} \tag{5.4}$$

where A_S is the cross-sectional area of the smaller piston.

Since the pressure is transmitted through the liquid to the second cylinder (Pascal's law), the force on the larger piston (F_L) is given by

$$F_L = pA_L \tag{5.5}$$

where A_L is the cross-sectional area of the large piston (assuming the pistons are at the same level), from which

$$F_L = \frac{A_L F_S}{A_S} \tag{5.6}$$

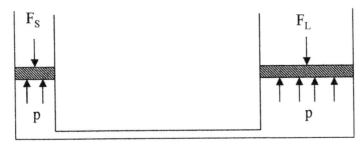

Figure 5.4 Diagram of a hydraulic press.

It can be seen that the force F_L is magnified by the ratio of the piston areas. This principle is used extensively in hoists, hydraulic equipment, and the like.

Example 5.9 In Fig. 5.4, if the area of the small piston A_S is 0.3 m² and the area of the large piston A_L is 5 m², what is the force F_L on the large piston, if the force F_S on the small piston is 85 N?

$$\text{Force } F_L \text{ on piston} = \frac{5 \text{ N}}{0.3 \times 85} = 1416.7 \text{ N}$$

A vacuum is very difficult to achieve in practice. Vacuum pumps can only approach a true vacuum. Good small volume vacuums, such as in a barometer, can be achieved. Pressures less than atmospheric pressure are often referred to as "negative gauge" and are indicated by an amount below atmospheric pressure, for example, 5 psig would correspond to 9.7 psia (assume atm = 14.7 psia).

5.5 Measuring Instruments

5.5.1 Manometers

Manometers are good examples of pressure measuring instruments, though they are not as common as they used to be because of the development of new, smaller, more rugged, and easier to use pressure sensors.

U–tube manometers consist of U-shaped glass tubes partially filled with a liquid. When there are equal pressures on both sides, the liquid levels will correspond to the zero point on a scale as shown in Fig. 5.5a. The scale is graduated in pressure units. When a pressure is applied to one side of the U-tube that is higher than on the other side, as shown in Fig. 5.5b, the liquid rises higher in the lower pressure side, so that the difference in the heights of the two columns of liquid compensates for the difference in pressure, as in Eq. (5.2). The pressure difference is given by

$$P_R - P_L = \gamma \times \text{difference in height of the liquid in the columns} \quad (5.7)$$

where γ is the specific weight of the liquid in the manometer.

Inclined manometers were developed to measure low pressures. The low-pressure arm is inclined, so that the fluid has a longer distance to travel than

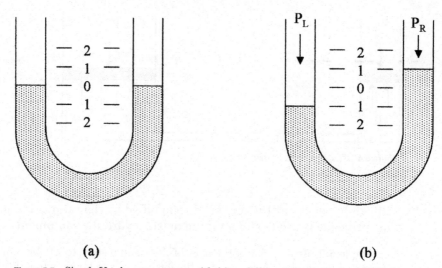

Figure 5.5 Simple U-tube manometers with (a) no differential pressure and (b) higher pressure on the left side.

in a vertical tube for the same pressure change. This gives a magnified scale as shown in Fig. 5.6a.

Well manometers are alternatives to inclined manometers for measuring low pressures using low-density liquids. In the well manometer, one leg has a much larger diameter than the other leg, as shown in Fig. 5.6b. When there is no pressure difference the liquid levels will be at the same height for a zero reading. An increase in the pressure in the larger leg will cause a larger change in the height of the liquid in the smaller leg. The pressure across the larger area of the well must be balanced by the same volume of liquid rising in the smaller leg. The effect is similar to the balance of pressure and volume in hydraulic jacks.

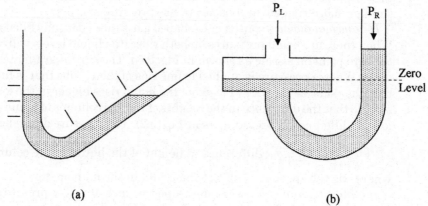

Figure 5.6 Other types of manometers are the (a) inclined-tube manometer and (b) well-type manometer.

Example 5.10 The liquid in a well manometer has a specific weight of 40 lb/ft³. How far will the liquid rise in the smaller leg, if the pressure in the larger leg is 1.5 lb/ft² higher than in the smaller leg?

$$h = \frac{p}{\gamma} = \frac{1.5 \text{ ft}}{40} = 0.45 \text{ in}$$

Example 5.11 The liquid in a manometer has a density of 850 kg/m³. What will be the difference in the liquid levels in the manometer tubes, if the differential pressure between the tubes is 5.2 kPa?

$$h = \frac{p}{\gamma} = \frac{5.2 \text{ kPa}}{850 \text{ kg/m}^3} \times \frac{\text{N/m}^2}{9.8 \text{ N/m}^3} = 62 \text{ cm}$$

5.5.2 Diaphragms, capsules, and bellows

Gauges are a major group of pressure sensors that measure pressure with respect to atmospheric pressure. Gauge sensors are usually devices that change their shape when pressure is applied. These devices include diaphragms, capsules, bellows, and Bourdon tubes.

A diaphragm consists of a thin layer or film of a material supported on a rigid frame and is shown in Fig. 5.8a. Pressure can be applied to one side of the film for gauge sensing or pressures can be applied to both sides of the film for differential or absolute pressure sensing. A wide range of materials can be used for the sensing film, from rubber to plastic for low-pressure devices, silicon for medium pressures, to stainless steel for high pressures. When pressure is applied to the diaphragm, the film distorts or becomes slightly spherical. This movement can be sensed using a strain gauge, piezoelectric, or changes in capacitance techniques (older techniques included magnetic and carbon pile devices). The deformation in the above sensing devices uses transducers to give electrical signals. Of all these devices the micromachined silicon diaphragm is the most commonly used industrial pressure sensor for the generation of electrical signals.

A silicon diaphragm uses silicon, which is a semiconductor. This allows a strain gauge and amplifier to be integrated into the top surface of the silicon structure after the diaphragm was etched from the back side. These devices have built-in temperature-compensated piezoelectric strain gauge and amplifiers that give a high output voltage (5 V FSD [volt full scale reading or deflection]). They are very small, accurate (2 percent FSD), reliable, have a good temperature operating range, are low cost, can withstand high overloads, have good longevity, and are unaffected by many chemicals. Commercially made devices are available for gauge, differential, and absolute pressure sensing up to 200 psi (1.5 MPa). This range can be extended by the use of stainless steel diaphragms to 100,000 psi (700 MPa).

Figure 5.7a shows the cross sections of the three configurations of the silicon chips (sensor dies) used in microminiature pressure sensors, i.e., gauge, absolute, and differential. The given dimensions illustrate that the sensing elements are very small. The die is packaged into a plastic case (about 0.2 in thick × 0.6 in diameter). A gauge assembly is shown in Fig. 5.7b. The sensor is used in blood pressure

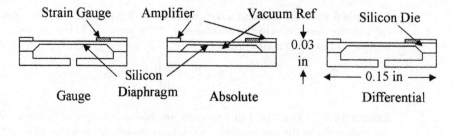

(a)

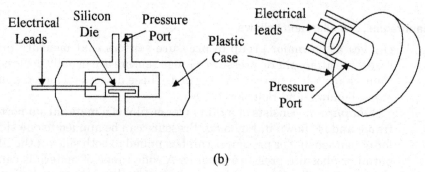

(b)

Figure 5.7 Cross section of (a) various types of microminiature silicon pressure sensor die and (b) a packaged microminiature gauge sensor.

monitors and many industrial applications, and is extensively used in automotive pressure-sensing applications, i.e., manifold air pressure, barometric air pressure, oil, transmission fluid, break fluid, power steering, tire pressure and the like.

Capsules are two diaphragms joined back to back, as shown in Fig. 5.8b. Pressure can be applied to the space between the diaphragms forcing them

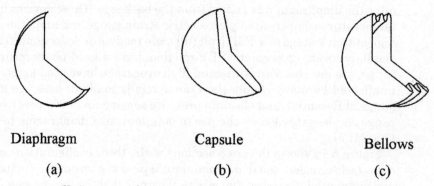

Figure 5.8 Various types of pressure-sensing elements: (a) diaphragm, (b) capsule, and (c) bellows.

apart to measure gauge pressure. The expansion of the diaphragm can be mechanically coupled to an indicating device. The deflection in a capsule depends on its diameter, material thickness, and elasticity. Materials used are phosphor bronze, stainless steel, and iron nickel alloys. The pressure range of instruments using these materials is up to 50 psi (350 kPa). Capsules can be joined together to increase sensitivity and mechanical movement.

Bellows are similar to capsules, except that the diaphragms instead of being joined directly together, are separated by a corrugated tube or tube with convolutions, as shown in Fig. 5.8c. When pressure is applied to the bellows it elongates by stretching the convolutions and not the end diaphragms. The materials used for the bellows type of pressure sensor are similar to those used for the capsule, giving a pressure range for the bellows of up to 800 psi (5 MPa). Bellows devices can be used for absolute and differential pressure measurements.

Differential measurements can be made by connecting two bellows mechanically, opposing each other when pressure is applied to them, as shown in Fig. 5.9a. When pressures P_1 and P_2 are applied to the bellows a differential scale reading is obtained. Figure 5.9b shows a bellows configured as a differential pressure transducer driving a *linear variable differential transformer* (LVDT) to obtain an electrical signal, P_2 could be the atmospheric pressure for gauge measurements. The bellows is the most sensitive of the mechanical devices for low-pressure measurements, i.e., 0 to 210 kPa.

5.5.3 Bourdon tubes

Bourdon tubes are hollow, cross-sectional beryllium, copper, or steel tubes, shaped into a three quarter circle, as shown in Fig. 5.10a. They may be rectangular or oval in cross section, but the operating principle is that the outer edge of the cross section has a larger surface than the inner portion. When pressure is applied, the outer edge has a proportionally larger total force applied because of its larger surface area, and the diameter of the circle increases. The walls of the tubes are between 0.01 and 0.05 in thick. The tubes are anchored at one end so that when

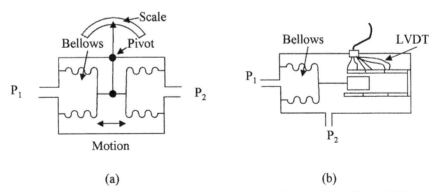

Figure 5.9 Differential bellows pressure gauges for (a) direct scale reading and (b) as a pressure transducer.

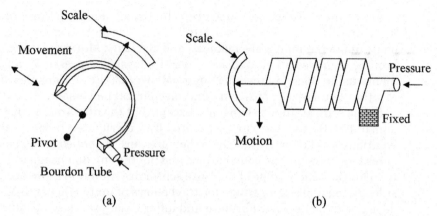

Figure 5.10 Pressure sensors shown are (a) the Bourdon tube and (b) the helical Bourdon tube.

pressure is applied to the tube, it tries to straighten and in doing so the free end of the tube moves. This movement can be mechanically coupled to a pointer, which when calibrated, will indicate pressure as a line of sight indicator, or it can be coupled to a potentiometer to give a resistance value proportional to the pressure for electrical signals. Figure 5.10b shows a helical pressure tube. This configuration is more sensitive than the circular Bourdon tube. The Bourdon tube dates from the 1840s. It is reliable, inexpensive, and one of the most common general purpose pressure gauges.

Bourdon tubes can withstand overloads of up to 30 to 40 percent of their maximum rated load without damage, but if overloaded may require recalibration. The tubes can also be shaped into helical or spiral shapes to increase their range. The Bourdon tube is normally used for measuring positive gauge pressures, but can also be used to measure negative gauge pressures. If the pressure on the Bourdon tube is lowered, then the diameter of the tube reduces. This movement can be coupled to a pointer to make a vacuum gauge. Bourdon tubes can have a pressure range of up to 100,000 psi (700 MPa). Figure 5.11 shows

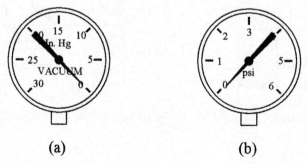

Figure 5.11 Bourdon-tube type pressure gauges for (a) negative and (b) positive pressures.

the Bourdon-tube type of pressure gauge when used for measuring negative pressure (vacuum) (*a*) and positive pressure (*b*). Note the counterclockwise movement in (*a*) and the clockwise movement in (*b*).

5.5.4 Other pressure sensors

Barometers are used for measuring atmospheric pressure. A simple barometer is the mercury barometer shown in Fig. 5.12*a*. It is now rarely used due to its fragility and the toxicity of mercury. The aneroid (no fluid) barometer is favored for direct reading (bellows in Fig. 5.9 or helical Bourdon tube in Fig. 5.10*b*) and the solid-state absolute pressure sensor for electrical outputs.

A piezoelectric pressure gauge is shown in Fig. 5.12*b*. Piezoelectric crystals produce a voltage between their opposite faces when a force or pressure is applied to the crystal. This voltage can be amplified and the device used as a pressure sensor.

Capacitive devices use the change in capacitance between the sensing diaphragm and a fixed plate to measure pressure. Some micromachined silicon pressure sensors use this technique in preference to a strain gauge. This technique is also used in a number of other devices to accurately measure any small changes in diaphragm deformation.

5.5.5 Vacuum instruments

Vacuum instruments are used to measure pressures less than atmospheric pressure. The Bourdon tube, diaphragms, and bellows can be used as vacuum gauges, but measure negative pressures with respect to atmospheric pressure. The silicon absolute pressure gauge has a built-in low-pressure reference, so it is calibrated to measure absolute pressures. Conventional devices can be used down to 20 torr (5 kPa). The range can be extended down to about 1 torr with special sensing devices.

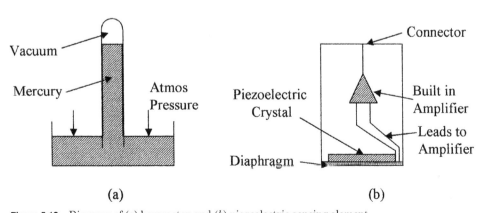

Figure 5.12 Diagram of (*a*) barometer and (*b*) piezoelectric sensing element.

The *Pirani gauge* and special setups using thermocouples can measure vacuums down to about 5 torr. These methods are based on the relation of heat conduction and radiation from a heating element to the number of gas molecules per unit volume in the low-pressure region, which determines the pressure.

Ionization gauges can be used to measure pressures down to about 2 torr. The gas is ionized with a beam of electrons and the current is measured between two electrodes in the gas. The current is proportional to the number of ions per unit volume, which is also proportional to the gas pressure.

McLeod gauge is a device set up to measure very low pressures, i. e., from 1 to 50 torr. The device compresses the low-pressure gas so that the increased pressure can be measured. The change in volume and pressure can then be used to calculate the original gas pressure, providing that the gas does not condense.

5.6 Application Considerations

When installing pressure sensors, care should be taken to select the correct pressure sensor for the application.

5.6.1 Selection

Pressure-sensing devices are chosen for pressure range, overload requirements, accuracy, temperature operating range, line-of-sight reading, or electrical signal, and response time. In some applications there are other special requirements. Parameters, such as hystersis and stability, should be obtained from the manufacturers' specifications. For most industrial applications reading positive pressures, the Bourdon tube is a good choice for direct visual readings and the silicon pressure sensor for the generation of electrical signals. Both types of devices have commercially available sensors to measure from a few pounds per square inch pressure FSD up to 10,000 psi (700 MPa) FSD. Table 5.3 gives a comparison of the two types of devices.

Table 5.4 lists the operating range for several types of pressure sensors.

5.6.2 Installation

The following should be taken into consideration when installing pressure-sensing devices.

1. Distance between sensor and source should be kept to a minimum.
2. Sensors should be connected via valves for ease of replacement.

TABLE 5.3 Comparison of Bourdon Tube Sensor and Silicon Sensor

Device	Maximum pressure range, lb/in^2	Accuracy FSD, %	Response time, s	Overload, %
Bourdon tube	10,000	2	1	40
Silicon sensor	10,000	2	1×10^{-3}	400

TABLE 5.4 Approximate Pressure Ranges for Pressure-Sensing Devices

Device	Maximum range, lb/in²	Device	Maximum range, lb/in²
U-tube manometer	15	Diaphragm	400
Bellows	800	Capsule	50
Bourdon tube	100,000	Spiral bourdon	40,000
Helical bourdon	80,000	Piezoelectric	100,000
Strain gauge	100,000	Solid state diaphragm	200
Stainless steel diaphragm	100,000		

3. Overrange protection devices should be included at the sensor.
4. To eliminate errors due to trapped gas in sensing liquid pressures, the sensor should be located below the source.
5. To eliminate errors due to trapped liquid in sensing gas pressures, the sensor should be located above the source.
6. When measuring pressures in corrosive fluids and gases, an inert medium is necessary between the sensor and the source or the sensor must be corrosion resistant.
7. The weight of the liquid in the connection line of a liquid pressure sensing device located above or below the source will cause errors in the zero, and a correction must be made by the zero adjustment, or otherwise compensated for in measurement systems.
8. Resistance and capacitance can be added to electron circuits to reduce pressure fluctuations and unstable readings.

5.6.3 Calibration

Pressure-sensing devices are calibrated at the factory. In cases where a sensor is suspect and needs to be recalibrated, the sensor can be returned to the factory for recalibration, or it can be compared to a known reference. Low-pressure devices can be calibrated against a liquid manometer. High-pressure devices can be calibrated with a dead-weight tester. In a dead-weight tester the pressure to the device under test is created by weights on a piston. High pressures can be accurately reproduced.

Summary

Pressure measurement standards in both English and SI units were discussed in this section. Pressure formulas and the types of instruments and sensors used in pressure measurements were given.

The main points discussed were:

1. Definitions of the terms and standards used in pressure measurements, both gauge and absolute pressures.
2. English and SI pressure measurement units and the relation between the two as well as atmospheric, torr, and millibar standards.
3. Pressure laws and formulas used in hydrostatic pressure measurements and buoyancy together with examples.
4. The various types of instruments including manometers, diaphragms, and micromachined pressure sensors. Various configurations for use in absolute and differential pressure sensing in both liquid and gas pressure measurements.
5. In the application section, the characteristics of pressure sensors were compared, and the considerations that should be made when installing pressure sensors were given.

Problems

5.1 A tank is filled with pure water. If the pressure at the bottom of the tank is 17.63 psig, what is the depth of the water?

5.2 What is the pressure on an object at the bottom of a fresh water lake if the lake is 123 m deep?

5.3 An instrument reads 1038 psf. If the instrument was calibrated in kilopascals, what would it read?

5.4 What will be the reading of a mercury barometer in centimeters if the atmospheric pressure is 14.75 psi?

5.5 A tank 2.2 ft × 3.1 ft × 1.79 ft weighs 1003 lb when filled with a liquid. What is the specific gravity of the liquid if the empty tank weighs 173 lb?

5.6 An open tank 3.2 m wide by 4.7 m long is filled to a depth of 5.7 m with a liquid whose SG is 0.83. What is the absolute pressure on the bottom of the tank in kilopascals?

5.7 Two pistons connected by a pipe are filled with oil. The larger piston has 3.2 ft diameter and has a force of 763 lb applied to it. What is the diameter of the smaller piston if it can support a force of 27 lb?

5.8 A block of wood with a density of 35.3 lb/ft^3 floats in a liquid with three-fourths of its volume submersed. What is the specific gravity of the liquid?

5.9 A 15.5-kg mass of copper has an apparent mass of 8.7 kg in oil whose SG is 0.77. What is the volume of the copper and its specific weight?

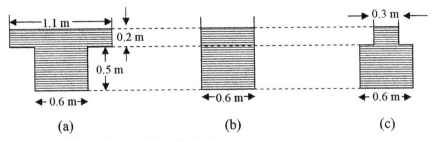

Figure 5.13 Figure for use with Prob. 5.20.

5.10 A dam is 283 m high when it is full of water. What is the pounds per square inch absolute at the bottom of the reservoir?

5.11 A liquid has a SG of 7.38. What is its specific weight in pound per cube foot and kilogram per cubic meter?

5.12 What is the equivalent of 25, 49, and 83 kPa in pounds per square inch?

5.13 The cabin pressure in a spacecraft is maintained at 14.3 psia. What will be the force on a window 2.9 ft wide and 1.7 ft high when the craft is in outer space?

5.14 A U-tube manometer uses glycerin as the measuring fluid. What will be the differential pressure if the distance between the levels of glycerin is 103 in?

5.15 An open tank contains 1.9 m of water floating on 10.3 cm of mercury. What is the pressure in pounds per square foot absolute on the bottom of the tank?

5.16 Oil (SG = 0.93) is pumped from a well. If the pump is 11.7 ft above the surface of the oil, what pressure must the pump be able to generate to lift the oil up to the pump?

5.17 A piston 8.7-in diameter has a pressure of 3.7 kPa on its surface. What force in SI units is applied to the piston?

5.18 The water pressure at the base of a water tower is 107.5 psi. What is the head of water?

5.19 A U-tube manometer reads a pressure of 270 torr. What is the pressure in pounds per square inch absolute?

5.20 Each of the three circular containers in Fig.5.13 contain a liquid with a SG of 1.37. What is the pressure in pascal gauge acting on the base of each container and the weight of liquid in each container?

Chapter 6

Level

Chapter Objectives

This chapter will help you understand the units used in level measurements and become familiar with the most common methods of using the various level standards.

Topics discussed in this chapter are as follows:

- The formulas used in level measurements
- The difference between direct and indirect level measuring devices
- The difference between continuous and single-point measurements
- The various types of instruments available for level measurements
- Application of the various types of level sensing devices

Most industrial processes use liquids such as water, chemicals, fuel, and the like, as well as free flowing solids (powders and granular materials). These materials are stored in containers ready for on-demand use. It is, however, imperative to know the levels and remaining volumes of these materials so that the containers can be replenished on an as needed basis to avoid the cost of large volume storage.

6.1 Introduction

This chapter discusses the measurement of the level of liquids and free flowing solids in containers. The detector is normally sensing the interface between a liquid and a gas, a solid and a gas, a solid and a liquid, or possibly the interface between two liquids. Sensing liquid levels fall into two categories; firstly, single-point sensing and secondly, continuous level monitoring. In the case of single-point sensing the actual level of the material is detected when it reaches a predetermined level, so that the appropriate action can be taken to prevent overflowing or to refill the container.

Continuous level monitoring measures the level of the liquid on an uninterrupted basis. In this case the level of the material will be constantly monitored and hence, the volume can be calculated if the cross-sectional area of the container is known.

Level measurements can be direct or indirect; examples of these are using a float technique or measuring pressure and calculating the liquid level. Free flowing solids are dry powders, crystals, rice, grain and so forth.

6.2 Level Formulas

Pressure is often used as an indirect method of measuring liquid levels. Pressure increases as the depth increases in a fluid. The pressure is given by

$$\Delta p = \gamma \Delta h \tag{6.1}$$

where Δp = change in pressure
γ = specific weight
Δh = depth

Note the units must be consistent, i.e., pounds and feet, or newtons and meters.

Buoyancy is an indirect method used to measure liquid levels. The level is determined using the buoyancy of an object partially immersed in a liquid. The buoyancy B or upward force on a body in a liquid can be calculated from the equation

$$B = \gamma \times \text{area} \times d \tag{6.2}$$

where area is the cross-sectional area of the object and d is the immersed depth of the object.

The liquid level is then calculated from the weight of a body in a liquid W_L, which is equal to its weight in air $(W_A - B)$, from which we get

$$d = \frac{W_A - W_L}{\gamma \times \text{area}} \tag{6.3}$$

The weight of a container can be used to calculate the level of the material in the container. In Fig. 6.1a the volume V of the material in the container is given by

$$V = \text{area} \times \text{depth} = \pi r^2 \times d \tag{6.4}$$

where r is the radius of the container and d is the depth of the material.

The weight of material W in a container is given by

$$W = \gamma V \tag{6.5}$$

Capacitive probes can be used in nonconductive liquids and free flowing solids for level measurement. Many materials, when placed between the plates of a capacitor, increase the capacitance by a factor μ called the dielectric constant of the material. For instance, air has a dielectric constant of 1 and water 80. Figure 6.1b shows two capacitor plates partially immersed in a nonconductive

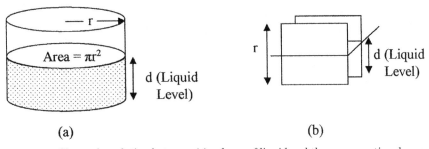

Figure 6.1 Shows the relation between (a) volume of liquid and the cross-sectional area and the liquid depth and (b) liquid level, plate capacitance, and a known dielectric constant in a nonconducting liquid.

liquid. The capacitance (Cd) is given by

$$Cd = Ca\mu\frac{d}{r} + Ca \qquad (6.6)$$

where Ca = capacitance with no liquid
μ = dielectric constant of the liquid between the plates
r = height of the plates
d = depth or level of the liquid between the plates

The dielectric constants of some common liquids are given in Table 6.1; there are large variations in dielectric constant with temperature so that temperature correction may be needed. In Eq. (6.6) the liquid level is given by

$$d = \frac{(Cd - Ca)}{\mu Ca} r \qquad (6.7)$$

6.3 Level Sensing Devices

There are two categories of level sensing devices. They are direct sensing, in which case the actual level is monitored, and indirect sensing where a property of the liquid such as pressure is sensed to determine the liquid level.

TABLE 6.1 Dielectric Constant of Some Common Liquids

Liquid	Dielectric constant
Water	80 @ 20°C
	88 @ 0°C
Glycerol	42.5 @ 25°C
	47.2 @ 0°C
Acetone	20.7 @ 25°C
Alcohol (Ethyl)	24.7 @ 25°C
Gasoline	2.0 @ 20°C
Kerosene	1.8 @ 20°C

6.3.1 Direct level sensing

Sight glass (open end/differential) or gauge is the simplest method for direct visual reading. As shown in Fig. 6.2 the sight glass is normally mounted vertically adjacent to the container. The liquid level can then be observed directly in the sight glass. The container in Fig. 6.2a is closed. In this case the ends of the glass are connected to the top and bottom of the tank, as would be used with a pressurized container (boiler) or a container with volatile, flammable, hazardous, or pure liquids. In cases where the tank contains inert liquids such as water and pressurization is not required, the tank and sight glass can both be open to the atmosphere as shown in Fig. 6.2b. The top of the sight glass must have the same pressure conditions as the top of the liquid or the liquid levels in the tank and sight glass will be different. In cases where the sight glass is excessively long, a second inert liquid with higher density than the liquid in the container can be used in the sight glass (see Fig. 6.2c). Allowance must be made for the difference in the density of the liquids. If the glass is stained or reacts with the containerized liquid the same approach can be taken or a different material can be used for the sight glass. Magnetic floats can also be used in the sight glass so that the liquid level can be monitored with a magnetic sensor such as a Hall effect device.

Floats (angular arm or pulley) are shown in Fig. 6.3. The figure shows two types of simple float sensors. The float material is less dense than the density of the liquid and floats up and down on top of the material being measured. In Fig. 6.3a a float with a pulley is used; this method can be used with either liquids or free flowing solids. With free flowing solids, agitation is sometimes used to level the solids. An advantage of the float sensor is that it is almost independent of the density of the liquid or solid being monitored. If the surface of the material being monitored is turbulent, causing the float reading to vary excessively, some means of damping might be used in the system. In Fig. 6.3b a ball float is attached to an arm; the angle of the arm is measured to indicate the level of the material (an example of the use of this type of sensor is the monitoring of the fuel level in the tank of an automobile). Although very simple and cheap to manufacture, the disadvantage of this type of float is its

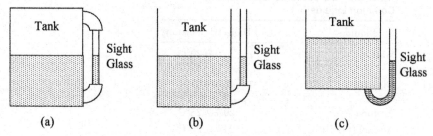

Figure 6.2 Various configurations of a sight glass to observe liquid levels (a) pressurized or closed container, (b) open container, and (c) higher density sight glass liquid.

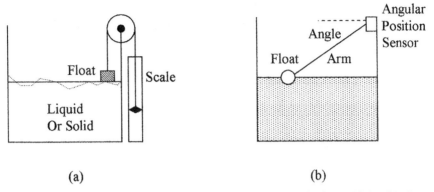

Figure 6.3 Methods of measuring liquid levels using (a) a simple float with level indicator on the outside of the tank and (b) an angular arm float.

nonlinearity as shown by the line of sight scale in Fig. 6.4a. The scale can be replaced with a potentiometer to obtain an electrical signal that can be linearized for industrial use.

Figure 6.4b shows an alternative method of using pulleys to obtain a direct visual scale that can be replaced by a potentiometer to obtain a linear electrical output with level.

A displacer with force sensing is shown in Fig. 6.5a. This device uses the change in the buoyant force on an object to measure the changes in liquid level. The displacers must have a higher specific weight than that of the liquid level being measured and have to be calibrated for the specific weight of the liquid. A force or strain gauge measures the excess weight of the displacer. There is only a small movement in this type of sensor compared to a float sensor.

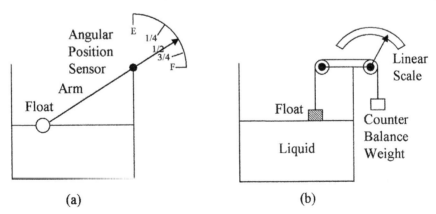

Figure 6.4 Scales used with float level sensors (a) nonlinear scale with angular arm float and (b) linear scale with a pulley type of float.

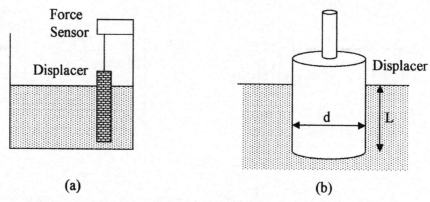

Figure 6.5 Displacer with a force sensor for measuring liquid level by (*a*) observing the loss of weight of the displacer due to the buoyancy forces of the displaced liquid and (*b*) dispenser dimensions.

The buoyant force on a cylindrical displacer shown in Fig. 6.5*b* using Eq. (6.2) is given by

$$F = \frac{\gamma \pi d^2 L}{4} \tag{6.8}$$

where γ = specific weight of the liquid
d = float diameter
L = length of the displacer submerged in the liquid

The weight as seen by the force sensor is given by

$$\text{Weight on force sensor} = \text{weight of displacer} - F \tag{6.9}$$

It should be noted that the units must be in the same measurement system and the liquid must not rise above the top of the displacer or the displacer must not touch the bottom of the container.

Example 6.1 A displacer with a diameter of 8 in is used to measure changes in water level. If the water level changes by 1 ft what is the change in force sensed by the force sensor?
From Eq. (6.9)

$$\text{Change in force} = (\text{weight of dispenser} - F_1) - (\text{weight of dispenser} - F_2)$$
$$= F_2 - F_1$$

From Eq. (6.8)

$$F_2 - F_1 = \frac{62.4 \text{ lb/ft}^3 \times \pi (8 \text{ ft})^2 \times 12^2}{4} = 21.8 \text{ lb}$$

Example 6.2 A 3.5-cm diameter displacer is used to measure acetone levels. What is the change in force sensed if the liquid level changes by 52 cm?

$$F_2 - F_1 = \frac{7.74 \text{ kN/m}^3 \times \pi 3.5^2 \text{ cm}^2 \times 52 \text{ cm}}{4 \times 10^6 \text{cm/m}} = 3.87 \text{ N}(395 \text{ g})$$

Probes for measuring liquid levels fall into three categories, i.e., conductive, capacitive, and ultrasonic.

Conductive probes are used for single-point measurements in liquids that are conductive and nonvolatile as a spark can occur. Conductive probes are shown in Fig. 6.6a. Two or more probes as shown can be used to indicate set levels. If the liquid is in a metal container, the container can be used as the common probe. When the liquid is in contact with two probes the voltage between the probes causes a current to flow indicating that a set level has been reached. Thus, probes can be used to indicate when the liquid level is low and to operate a pump to fill the container. Another or a third probe can be used to indicate when the tank is full and to turn off the filling pump.

Capacitive probes are used in liquids that are nonconductive and have a high μ and can be used for continuous level monitoring. The capacitive probe shown in Fig. 6.6b consists of an inner rod with an outer shell; the capacitance is measured between the two using a capacitance bridge. In the portion out of the liquid, air serves as the dielectric between the rod and outer shell. In the immersed section, the dielectric is that of the liquid that causes a large capacitive change, if the tank is made of metal it can serve as the outer shell. The capacitance change is directly proportional to the level of the liquid. The dielectric constant of the liquid must be known for this type of measurement. The dielectric constant can vary with temperature so that temperature correction may be required.

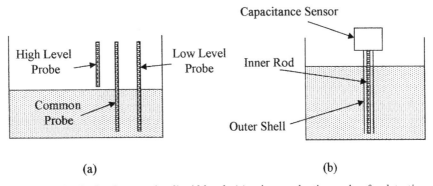

Figure 6.6 Methods of measuring liquid levels (a) using conductive probes for detecting set levels and (b) a capacitive probe for continuous monitoring.

Example 6.3 A capacitive probe 30-in long has a capacitance of 22 pF in air. When partially immersed in water with a dielectric constant of 80 the capacitance is 1.1 nF. What is the length of the probe immersed in water?

From Eq. (6.6)

$$d = \frac{(1.1 \times 10^3 \text{ pf} - 22 \text{ pf})30 \text{ in}}{80 \times 22 \text{ pf}} = 18.4 \text{ in}$$

Ultrasonics can be used for single point or continuous level measurement of a liquid or a solid. A single ultrasonic transmitter and receiver can be arranged with a gap as shown in Fig. 6.7a to give single-point measurement. As soon as liquid fills the gap, ultrasonic waves from the transmitter reach the receiver. A setup for continuous measurement is shown in Fig. 6.7b. Ultrasonic waves from the transmitter are reflected by the surface of the liquid to the receiver; the time for the waves to reach the receiver is measured. The time delay gives the distance from the transmitter and receiver to the surface of the liquid, from which the liquid level can be calculated knowing the velocity of ultrasonic waves. As there is no contact with the liquid, this method can be used for solids and corrosive and volatile liquids. In a liquid the transmitter and receiver can also be placed on the bottom of the container and the time measured for a signal to be reflected from the surface of the liquid to the receiver to measure the depth of the liquid.

6.3.2 Indirect level sensing

The most commonly used method of indirectly measuring a liquid level is to measure the hydrostatic pressure at the bottom of the container. The depth can then be extrapolated from the pressure and the specific weight of the liquid can be calculated using Eq. (6.1). The pressure can be measured by any of the methods given in the section on pressure. The dial on the pressure gauge can be calibrated directly in liquid depth. The depth of liquid can also be measured using bubblers, radiation, resistive tapes, and by weight measurements.

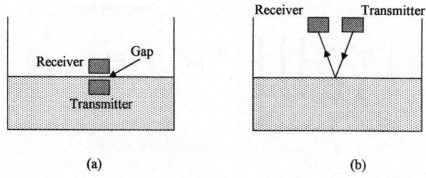

Figure 6.7 Use of ultrasonics for (a) single-point liquid level measurement and (b) continuous liquid level measurements made by timing reflections from the surface of the liquid.

Example 6.4 A pressure gauge located at the base of an open tank containing a liquid with a specific weight of 54.5 lb/ft^3 registers 11.7 psi. What is the depth of the fluid in the tank?

From Eq. (6.1)

$$h = \frac{p}{\gamma} = \frac{11.7 \text{ psi} \times 144}{54.5 \text{ lb/ft}^3} = 30.9 \text{ ft}$$

Bubbler devices require a supply of clean air or inert gas. The setup is shown in Fig. 6.8a. Gas is forced through a tube whose open end is close to the bottom of the tank. The specific weight of the gas is negligible compared to the liquid and can be ignored. The pressure required to force the liquid out of the tube is equal to the pressure at the end of the tube due to the liquid, which is the depth of the liquid multiplied by the specific weight of the liquid. This method can be used with corrosive liquids as the material of the tube can be chosen to be corrosion resistant.

Example 6.5 How far below the surface of the water is the end of a bubbler tube, if bubbles start to emerge from the end of the tube when the air pressure in the bubbler is 148 kPa?

From Eq. (6.1)

$$h = \frac{p}{\gamma} = \frac{148 \text{ kPa} \times 10^{-4}}{1 \text{ gm/cm}^3} = 14.8 \text{ cm}$$

Radiation methods are sometimes used in cases where the liquid is corrosive, very hot, or detrimental to installing sensors. For single-point measurement only one transmitter and a detector are required. If several single-point levels are required, a detector will be required for each level measurement as shown in Fig. 6.8b. The disadvantages of this system are the cost and the need to handle radioactive material.

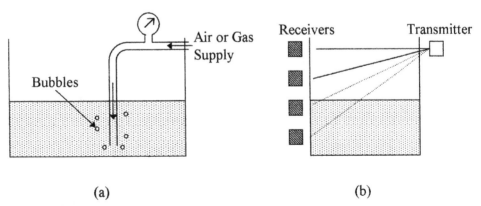

Figure 6.8 Liquid level measurements can be made (a) using a bubbler technique or (b) using a radiation technique.

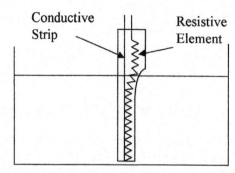

Figure 6.9 Demonstrating a resistive tape level sensor.

Resistive tapes can be used to measure liquid levels (see Fig. 6.9). A resistive element is placed in close proximity to a conductive strip in an easily compressible nonconductive sheath; the pressure of the liquid pushes the resistive element against the conductive strip, shorting out a length of the resistive element proportional to the depth of the liquid. The sensor can be used in liquids or slurries, it is cheap but is not rugged or accurate, it is prone to humidity problems, and measurement accuracy depends on material density.

Load cells can be used to measure the weight of a tank and its contents. The weight of the container is subtracted from the reading, leaving the weight of the contents of the container. Knowing the cross-sectional area of the tank and the specific weight of the material, the volume and/or depth of the contents can be calculated. This method is well suited for continuous measurement and the material being weighed does not come into contact with the sensor. Figure 6.10 shows two elements that can be used in load sensors for measuring force. Figure 6.10a shows a cantilever beam used as a force or weight sensor. The beam is rigidly attached at one end and a force is applied to the other end, a strain gauge attached to the beam is used to measure the strain in the beam, a second

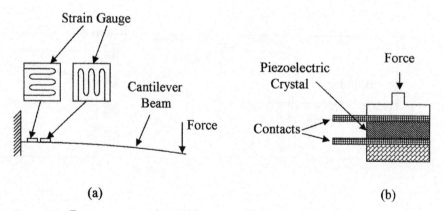

Figure 6.10 Force sensors can be used for measuring weight using (*a*) strain gauge technique or (*b*) a piezoelectric technique.

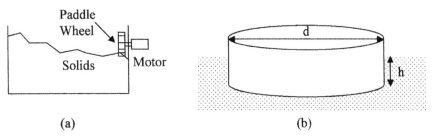

Figure 6.11 Shows (a) Paddle wheel technique to measure the level of free flowing solids and (b) a typical float shape.

strain gauge is used for temperature compensation. Figure 6.10b shows a piezoelectric sensor used to measure force or weight. The sensor gives an output voltage proportional to the force applied.

Example 6.6 What is the depth of the liquid in a container, if the specific weight of the liquid is 82 lb/ft^3; the container weights 45 lb and is 21 in in diameter? A load cell measures a total weight of 385 lb.

Using Eq. (6.4) and (6.5) we get the following:

$$\text{Weight of liquid} = 385 - 45 = 340 \text{ lb}$$

$$\text{Volume of liquid} = \frac{3.14 \times 21 \times 21 \times d}{12 \times 12 \times 4} \text{ ft} = \frac{340 \text{ lb}}{82 \text{ lb/ft}^3}$$

$$\text{Depth } (d) = \frac{4.15 \times 576}{1384.7} \text{ ft} = 1.73 \text{ ft} = 20.7 \text{ in}$$

Paddle wheels driven by electric motors can be used for sensing the level of solids in the form of power, grains, or granules. When the material reaches and covers the paddle wheel, the torque needed to turn the motor greatly increases. The torque can be an indication of the depth of the material. Such a set up is shown in Fig. 6.11a. Some agitation may be required to level the solid particles.

6.4 Application Considerations

A number of factors affect the choice of sensor for level measurement, such as pressure on the liquid, liquid temperature, turbulence, volatility, corrosiveness, accuracy needed, single-point or continuous measurement, direct or indirect, particulates in a liquid, free flowing solids, and so forth.

Floats are often used to sense fluid levels because they are unaffected by particulates, can be used for slurries, can be used with a wide range of liquid specific weights, and flat floats due to their area are less susceptible to turbulence on the surface of the liquid. Figure 6.11b shows a commonly used design for a float which can be attached to a level indicator. The float displaces its own weight of liquid as follows:

$$\text{Float weight} = \text{buoyant force} = \frac{\gamma_L \pi d^2 h}{4} \qquad (6.10)$$

where γ_L = specific weight of the liquid
d = diameter
h = immersion depth of the float

When the float is used to measure one or more feet of liquid depth, any change in h due to large changes in γ_L will have minimal effect on the measured liquid depth.

Displacers must never be completely submerged when measuring liquid depth and must have a specific weight greater than that of the liquid. Care must also be taken to ensure that the displacer is not corroded by the liquid and the specific weight of the liquid is constant over time. The temperature of the liquid may also have to be monitored to make corrections for density changes. Displacers can be used to measure depths up to about 3 m with an accuracy of ±0.5 cm.

Capacitive device accuracy can be affected by the placement of the device, so the manufacturer's installation instructions must be followed. The dielectric constant of the liquid should also be regularly monitored. Capacitive devices can be used in pressurized containers up to 30 MPa and temperatures up to 1000°C, and measure depths up to 6 m with an accuracy of ±1 percent.

Pressure gauge choice for measuring liquid levels can depend on a number of considerations, which are as follows:

1. The presence of particulates that can block the line to the gauge
2. Damage caused by excessive temperatures in the liquid
3. Damage due to peak pressure surges
4. Corrosion of the gauge by the liquid
5. Differential pressure gauges are needed if the liquid is under pressure
6. Distance between the tank and the gauge
7. Use of manual valves for gauge repair

Differential pressure gauges can be used in pressurized containers up to 30 MPa and temperatures up to 600°C to give accuracies of ±1 percent, the liquid depth depends on its density and the pressure gauge used.

Bubbler devices require certain precautions when being used. To ensure a continuous air or gas supply, the gas used must not react with the liquid. It may be necessary to install a one way valve to prevent the liquid being sucked back into the gas supply lines if the gas pressure is lost. The bubbler tube must be chosen so that it is not corroded by the liquid. Bubbler devices are typically used at atmospheric pressure, accuracies of about 2 percent can be obtained, depth depends on gas pressure available, and so forth.

Ultrasonic devices can be used with pressurized containers up to 2 MPa and 100°C temperature range for depths of up to 30 m with accuracies of about 2 percent.

Radiation devices are used for point measurement of hazardous materials. Due to the hazardous nature of the material, personnel should be trained in its use, transportation, storage, identification, and disposal.

Other considerations are that liquid level measurements can be effected by turbulence, readings may have to be averaged, and/or baffles used to reduce the turbulence. Frothing in the liquid can also be a source of error particularly with resistive or capacitive probes.

Summary

This chapter introduced the concepts of level measurement. The instruments used for direct and indirect measurement have been described and the application of level measuring instruments considered.

The key points covered in this chapter are as follows:

1. The formulas used by instruments for the measurement of liquid levels and free flowing solids with examples
2. The various types of instruments used to give direct measurement of liquid levels and the methods used to indirectly measure liquid levels
3. The difference between continuous and single-point level measurements in a liquid
4. Application considerations when selecting an instrument for measuring liquid and free flowing solid levels

Problems

6.1 What is the specific weight of a liquid, if the pressure is 4.7 psi at a depth of 17 ft?

6.2 What is the depth of a liquid, if the pressure is 127 kPa and the liquid density is 1.2 g/cm^3?

6.3 What is the displaced volume in cubic meters if the buoyancy on an object is 15 lb and the density of the liquid is 785 kg/m^3?

6.4 What is the liquid density in gram per cubic centimeter, if the buoyancy is 833 N on a 135 cm^3 submerged object?

6.5 The weight of a body in air is 17 lb and submerged in water is 3 lb. What is the volume and specific weight of the body?

6.6 A material has a density of 1263 kg/m^3. A block of the material weighs 72 kg when submerged in water. What is its volume and weight in air?

6.7 A container of 4.5-ft diameter is full of liquid. If the liquid has a specific weight of 63 lb/ft^3, what is the depth of the liquid if the weight of the container and liquid is 533 lb? Assume the container weighs 52 lb.

6.8 The weight of liquid in a round container is 1578 kg, the depth of the liquid is 3.2 m. If the density of the liquid is 0.83 g/cm^3, what is the diameter of the container?

6.9 A capacitive sensor is 3 ft 3 in high and has a capacitance of 25 pF in air and 283 pF when immersed in a liquid to a depth of 2 ft 7 in. What is the dielectric constant of the liquid?

6.10 A capacitive sensor 2.4 m in height has a capacitance of 75 pF in air if the sensor is placed in a liquid with a dielectric constant of 65 to a depth of 1.7 m. What will be the capacitive reading of the sensor?

6.11 A pressure gauge at the bottom of a tank reads 32 kPa. If the tank has 3.2 m diameter, what is the weight of liquid in the container?

6.12 What pounds per square inch is required by a bubbler system to produce bubbles at a depth of 4 ft 7 in water?

6.13 A bubbler system requires a pressure of 28 kPa to produce bubbles in a liquid with a density of 560 kg/m^3. What is the depth of the outlet of the bubbler in the liquid?

6.14 A displacer with a diameter of 4.7 cm is used to measure changes in the level of a liquid with a density of 470 kg/m^3. What is the change in force on the sensor if the liquid level changes 13.2 cm?

6.15 A displacer is used to measure changes in liquid level. The liquid has a density of 33 lb/ft^3. What is the diameter of the dispenser if a change in liquid level of 45 in produces a change in force on the sensor of 3.2 lb?

6.16 A bubbler system requires a pressure of 47 kPa to produce bubbles at a depth of 200 in. What is the density of the liquid in pounds per cubic foot?

6.17 A capacitive sensing probe 2.7 m high has a capacitance of 157 pF in air and 7.4 nF when partially immersed in a liquid with a dielectric constant of 79. How much of the probe is immersed in the liquid?

6.18 A force sensor is immersed in a liquid with a density of 61 lb/ft^3 to a depth of 42 in and then placed in a second liquid with a density of 732 N/m^3. What is the change in force on the sensor if the diameter of the sensor is 8 cm and the change in depth is 5.9 cm?

6.19 An ultrasonic transmitter and receiver are placed 10.5 ft above the surface of a liquid. How long will the sound waves take to travel from the transmitter to the receiver? Assume the velocity of sound waves is 340 m/s

6.20 If the liquid in Prob. 6.19 is lowered to 6.7 ft, what is the increase in time for the sound waves to go from the transmitter to the receiver?

Chapter 7

Flow

Chapter Objectives

This chapter will introduce you to the concepts of fluid velocity and flow and its relation to pressure and viscosity. The chapter will help you understand the units used in flow measurement and become familiar with the most commonly used flow standards.

This chapter covers the following topics:

- Reynolds number and its application to flow patterns
- Formulas used in flow measurements
- Bernoulli equation and its applications
- Difference between flow rate and total flow
- Pressure losses and their effects on flow
- Flow measurements using differential pressure measuring devices and their characteristics
- Open channel flow and its measurement
- Considerations in the use of flow instrumentation

7.1 Introduction

This chapter discusses the basic terms and formulas used in flow measurements and instrumentation. The measurement of fluid flow is very important in industrial applications. Optimum performance of some equipment and operations require specific flow rates. The cost of many liquids and gases are based on the measured flow through a pipeline making it necessary to accurately measure and control the rate of flow for accounting purposes.

7.2 Basic Terms

This chapter will be using terms and definitions from previous chapters as well as introducing a number of new definitions related to flow and flow rate sensing.

Velocity is a measure of speed and direction of an object. When related to fluids it is the rate of flow of fluid particles in a pipe. The speed of particles in a fluid flow varies across the flow, i.e., where the fluid is in contact with the constraining walls (the boundary layer) the velocity of the liquid particles is virtually zero; in the center of the flow the liquid particles will have the maximum velocity. Thus, the average rate of flow is used in flow calculations. The units of flow are normally feet per second (fps), feet per minute (fpm), meters per second (mps), and so on. Previously, the pressures associated with fluid flow were defined as static, impact, or dynamic.

Laminar flow of a liquid occurs when its average velocity is comparatively low and the fluid particles tend to move smoothly in layers, as shown in Fig. 7.1a. The velocity of the particles across the liquid takes a parabolic shape.

Turbulent flow occurs when the flow velocity is high and the particles no longer flow smoothly in layers and turbulence or a rolling effect occurs. This is shown in Fig. 7.1b. Note also the flattening of the velocity profile.

Viscosity is a property of a gas or liquid that is a measure of its resistance to motion or flow. A viscous liquid such as syrup has a much higher viscosity than water and water has a higher viscosity than air. Syrup, because of its high viscosity, flows very slowly and it is very hard to move an object through it. Viscosity (dynamic) can be measured in poise or centipoise, whereas kinematic viscosity (without force) is measured in stokes or centistokes. Dynamic or absolute viscosity is used in the Reynolds and flow equations. Table 7.1 gives a list of conversions. Typically the viscosity of a liquid decreases as temperature increases.

The *Reynolds number R* is a derived relationship combining the density and viscosity of a liquid with its velocity of flow and the cross-sectional dimensions

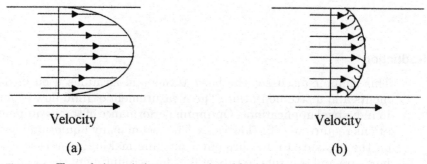

Figure 7.1 Flow velocity variations across a pipe with (a) laminar flow and (b) turbulent flow.

TABLE 7.1 Conversion Factors for Dynamic and Kinematic Viscosities

Dynamic viscosities	Kinematic viscosities
1 lb s/ft² = 47.9 Pa s	1 ft²/s = 9.29 × 10⁻² m²/s
1 centipoise = 10 Pa s	1 stoke = 10⁻⁴ m²/s
1 centipoise = 2.09 ×10⁻⁵ lb s/ft²	1 m²/s = 10.76 ft²/s
1 poise = 100 centipoise	1 stoke = 1.076 × 10⁻³ ft²/s

of the flow and takes the form

$$R = \frac{VD\rho}{\mu} \tag{7.1}$$

where V = average fluid velocity
D = diameter of the pipe
ρ = density of the liquid
μ = absolute viscosity

Flow patterns can be considered to be laminar, turbulent, or a combination of both. Osborne Reynolds observed in 1880 that the flow pattern could be predicted from physical properties of the liquid. If the Reynolds number for the flow in a pipe is equal to or less than 2000 the flow will be laminar. From 2000 to about 5000 is the intermediate region where the flow can be laminar, turbulent, or a mixture of both, depending upon other factors. Beyond 5000 the flow is always turbulent.

The *Bernoulli equation* is an equation for flow based on the law of conservation of energy, which states that the total energy of a fluid or gas at any one point in a flow is equal to the total energy at all other points in the flow.

Energy factors. Most flow equations are based on the law of energy conservation and relate the average fluid or gas velocity, pressure, and the height of fluid above a given reference point. This relationship is given by the Bernoulli equation. The equation can be modified to take into account energy losses due to friction and increase in energy as supplied by pumps.

Energy losses in flowing fluids are caused by friction between the fluid and the containment walls and by fluid impacting an object. In most cases these losses should be taken into account. Whilst these equations apply to both liquids and gases, they are more complicated in gases because of the fact that gases are compressible.

Flow rate is the volume of fluid passing a given point in a given amount of time and is typically measured in gallons per minute (gpm), cubic feet per minute (cfm), liter per minute, and so on. Table 7.2 gives the flow rate conversion factors.

Total flow is the volume of liquid flowing over a period of time and is measured in gallons, cubic feet, liters and so forth.

TABLE 7.2 Flow Rate Conversion Factors

1 gal/min = 6.309 ×10⁻⁵ m³/s	1 L/min = 16.67 × 10⁻⁶ m³/s
1 gal/min = 3.78 L/min	1 cu ft/sec = 449 gal/min
1 gal/min = 0.1337 ft³/min	1 gal/min = 0.00223 ft³/s
1 gal water = 231 in³	1 cu ft water = 7.48 gal

1 gal water = 0.1337 ft³ = 231 in³; 1 gal water = 8.35 lb; 1 ft³ water = 7.48 gal; 1000 liter water = 1 m³; 1 liter water = 1 kg

7.3 Flow Formulas

7.3.1 Continuity equation

The continuity equation states that if the overall flow rate in a system is not changing with time (see Fig. 7.2a), the flow rate in any part of the system is constant. From which we get the following equation:

$$Q = VA \tag{7.2}$$

where Q = flow rate
V = average velocity
A = cross-sectional area of the pipe

The units on both sides of the equation must be compatible, i.e., English units or metric units.

Example 7.1 What is the flow rate through a pipe 9 in diameter, if the average velocity is 5 fps?

$$Q = \frac{5 \text{ ft/s} \times \pi \times 0.75^2 \text{ ft}^2}{4} = 2.21 \text{ cfs} = \frac{2.21 \text{ gps}}{0.137} = 16.1 \text{ gps} = 16.1 \times 60 \text{ gpm} = 968 \text{ gpm}$$

If liquids are flowing in a tube with different cross section areas, i.e., A_1 and A_2, as is shown in Fig. 7.2b, the continuity equation gives

$$Q = V_1 A_1 = V_2 A_2 \tag{7.3}$$

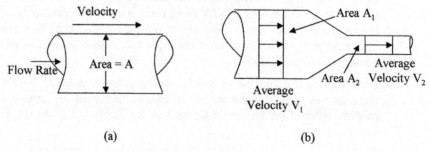

Figure 7.2 Flow diagram used for use in the continuity equation: (a) fixed diameter and (b) effects of different diameters on the flow rate.

Example 7.2 If a pipe goes from a 9-cm diameter to 6-cm diameter and the velocity in the 9-cm section is 2.21 m/s, what is the average velocity in the 6-cm section?

$$Q = V_1 A_1 = V_2 A_2$$

$$V_2 = \frac{2.21 \text{ m}^3/\text{s} \times \pi \times 4.5^2}{\pi \times 3^2} = 4.97 \text{ m/s}$$

Mass flow rate F is related to volume flow rate Q by

$$F = \rho Q \tag{7.4}$$

where F is the mass of liquid flowing and ρ is the density of the liquid.

As a gas is compressible, Eq. (7.3) must be modified for gas flow to

$$\gamma_1 V_1 A_1 = \gamma_2 V_2 A_2 \tag{7.5}$$

where γ_1 and γ_2 are specific weights of the gas in the two sections of pipe.

Equation (7.3) is the rate of weight flow in the case of a gas. However, this could also apply to liquid flow in Eq. (7.3) by multiplying both sides of the equation by the specific weight γ.

7.3.2 Bernoulli equation

The Bernoulli equation gives the relation between pressure, fluid velocity, and elevation in a flow system. The equation is accredited to Bernoulli (1738). When applied to Fig. 7.3a the following is obtained-

$$\frac{P_A}{\gamma_A} + \frac{V_A^2}{2g} + Z_A = \frac{P_B}{\gamma_B} + \frac{V_B^2}{2g} + Z_B \tag{7.6}$$

where P_A and P_B = absolute static pressures at points A and B, respectively
γ_A and γ_B = specific weights
V_A and V_B = average fluid velocities
g = acc of gravity
Z_A and Z_B = elevations above a given reference level, i.e., $Z_A - Z_B$ is the head of fluid.

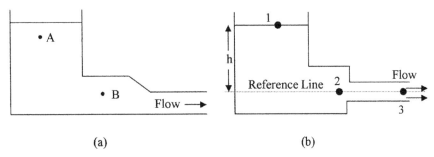

Figure 7.3 Container diagrams (a) the pressures at points A and B are related by the Bernoulli equation and (b) application of the Bernoulli in Example 7.3.

The units in Eq. (7.6) are consistent and reduce to units of length (feet in the English system and meter in the SI system of units) as follows:

$$\text{Pressure energy} = \frac{p}{\gamma} = \frac{\text{lb/ft}^2(\text{N/m}^2)}{\text{lb/ft}^3(\text{N/m}^3)} = \text{ft(m)}$$

$$\text{Kinetic energy} = \frac{V^2}{g} = \frac{(\text{ft/s})^2(\text{m/s})^2}{\text{ft/s}^2(\text{m/s}^2)} = \text{ft(m)}$$

$$\text{Potential energy} = Z = \text{ft(m)}$$

This equation is a conservation of energy equation and assumes no loss of energy between points A and B. The first term represents energy stored due to pressure, the second term represents kinetic energy or energy due to motion, and the third term represents potential energy or energy due to height. This energy relationship can be seen if each term is multiplied by mass per unit volume which cancels as the mass per unit volume is the same at points A and B. The equation can be used between any two positions in a flow system. The pressures used in the Bernoulli equation must be absolute pressures.

In the fluid system shown in Fig. 7.3b the flow velocity V at point 3 can be derived from Eq. (7.6) and is as follows using point 2 as the reference line.

$$\frac{P_1}{\gamma_1} + 0 + h = \frac{P_3}{\gamma_3} + \frac{V_3^2}{2g} + 0$$

$$V_3 = \sqrt{(2\,gh)} \tag{7.7}$$

Point 3 at the exit has dynamic pressure but no static pressure above 1 atm, and hence, $P_3 = P_1 = 1$ atm and $\gamma_1 = \gamma_3$. This shows that the velocity of the liquid flowing out of the system is directly proportional to the square root of the height of the liquid above the reference point.

Example 7.3 If h in Fig. 7.3b is 7.5 m, what is the pressure at P_2? Assume the areas at points 2 and 3 are 0.48 m^2 and 0.3 m^2 respectively.

$$V_3 = \sqrt{(2 \times 9.8 \times 7.5)} = 12.12 \text{ m/s}$$

Considering points 2 and 3 with the use of Eq. (7.6)

$$\frac{P_2}{9.8 \text{ kN}} + \frac{V_2^2}{2 \times 9.8} + 0 = \frac{101.3 \text{ kPa}}{9.8 \text{ kN}} + \frac{V_3^2}{2 \times 9.8} + 0 \tag{7.8}$$

Using the continuity Eq. (7.3) and knowing that the areas at point 2 and 3 are 0.48 m^2 and 0.3 m^2, respectively, the velocity at point 2 is given by-

$$V_2 = \left(\frac{A_3}{A_2}\right) V_3 = \left(\frac{0.3}{0.48}\right) 12.12 \text{ m/s} = 7.58 \text{ m/s}$$

Substituting the values obtained for V_2 and V_3 into Eq. (7.8), gives the following:

$$\frac{P_2}{9.8} + \frac{(7.58)^2}{2 \times 9.8} + 0 = \frac{101.3}{9.8} + \frac{(12.12)^2}{2 \times 9.8} + 0$$

$$P_2 = 146 \text{ kPa(a)} = 44.7 \text{ kPa (g)}$$

7.3.3 Flow losses

The Bernoulli equation does not take into account flow losses; these losses are accounted for by pressure losses and fall into two categories. Firstly, those associated with viscosity and the friction between the constriction walls and the flowing fluid, and secondly, those associated with fittings, such as valves, elbows, tees, and so forth.

Outlet losses. The flow rate Q from the continuity equation for point 3 in Fig. 7.3b for instance gives

$$Q = V_3 A_3$$

However, to account for losses at the outlet, the equation should be modified to

$$Q = C_D V_3 A_3 \qquad (7.9)$$

where C_D is the discharge coefficient that is dependent on the shape and size of the orifice. The discharge coefficients can be found in flow data handbooks.

Frictional losses. They are losses from liquid flow in a pipe due to friction between the flowing liquid and the restraining walls of the container. These frictional losses are given by

$$h_L = \frac{fLV^2}{2\,Dg} \qquad (7.10)$$

where h_L = head loss
f = friction factor
L = length of pipe
D = diameter of pipe
V = average fluid velocity
g = gravitation constant

The friction factor f depends on the Reynolds number for the flow and the roughness of the pipe walls.

Example 7.4 What is the head loss in a 2-in diameter pipe 120-ft long? The friction factor is 0.03 and the average velocity in the pipe is 11 fps.

$$h_L = \frac{fLV^2}{2\,Dg} = \frac{0.03 \times 120 \text{ ft} \times (11 \text{ ft/s})^2 12}{2 \text{ ft} \times 2 \times 32.2 \text{ ft/s}^2} = 40.6 \text{ ft}$$

This would be equivalent to $\dfrac{40.6 \text{ ft} \times 62.4 \text{ lb/ft}^3}{144} = 17.6$ psi

Fitting losses are losses due to couplings and fittings, which are normally less than those associated with friction and are given by

$$h_L = \frac{KV^2}{2g} \qquad (7.11)$$

where h_L = head loss due to fittings
K = loss coefficient for various fittings
V = average fluid velocity
g = gravitation constant

Values for K can be found in flow handbooks. Table 7.3 gives some typical values for the head loss coefficient factor in some common fittings.

Example 7.5 Fluid is flowing at 4.5 fps through 1 in fittings as follows: 5 × 90° ells, 3 tees, 1 gate valve, and 12 couplings. What is the head loss?

$$h_L = \frac{(5 \times 1.5 + 3 \times 0.8 + 1 \times 0.22 + 12 \times 0.085)\, 4.5 \times 4.5}{2 \times 32.2}$$

$$h_L = (7.5 + 2.4 + 0.22 + 1.02)\, 0.31 = 3.5 \text{ ft}$$

To take into account losses due to friction and fittings, the Bernouilli Eq. (7.6) is modified as follows:

$$\frac{P_A}{\gamma_A} + \frac{V_A^2}{2g} + Z_A = \frac{P_B}{\gamma_B} + \frac{V_B^2}{2g} + Z_B + h_{Lfriction} + h_{Lfittings} \qquad (7.12)$$

Form drag is the impact force exerted on devices protruding into a pipe due to fluid flow. The force depends on the shape of the insert and can be calculated from

$$F = C_D \gamma \frac{AV^2}{2g} \qquad (7.13)$$

where F = force on the object
C_D = drag coefficient
γ = specific weight
g = acceleration due to gravity
A = cross-sectional area of obstruction
V = average fluid velocity

TABLE 7.3 Typical Head Loss Coefficient Factors for Fittings

Threaded ell – 1 in	1.5	Flanged ell – 1 in	0.43
Threaded tee – 1 in inline	0.9	Branch	1.8
Globe valve (threaded)	8.5	Gauge valve (threaded)	0.22
Coupling or union – 1 in	0.085	Bell mouth reducer	0.05

TABLE 7.4 Typical Drag Coefficient Values for Objects Immersed in Flowing Fluid

Circular cylinder with axis perpendicular to flow	0.33 to 1.2
Circular cylinder with axis parallel to flow	0.85 to 1.12
Circular disk facing flow	1.12
Flat plate facing flow	1.9
Sphere	0.1 +

Flow handbooks contain drag coefficients for various objects. Table 7.4 gives some typical drag coefficients.

Example 7.6 A 5-in diameter ball is traveling through the air with a velocity of 110 fps, if the density of the air is 0.0765 lb/ft^3 and $C_D = 0.5$. What is the force acting on the ball?

$$F = C_D \gamma \frac{AV^2}{2g} = \frac{0.5 \times 0.0765 \text{ lb/ft}^3 \times \pi \times 5^2 \text{ ft}^2 \times (110 \text{ ft/s})^2}{2 \times 32.2 \text{ ft/s}^2 \times 4 \times 144} = 0.98 \text{ lb}$$

7.4 Flow Measurement Instruments

Flow measurements are normally indirect measurements using differential pressures to measure the flow rate. Flow measurements can be divided into the following groups: flow rate, total flow, and mass flow. The choice of the measuring device will depend on the required accuracy and fluid characteristics (gas, liquid, suspended particulates, temperature, viscosity, and so on.)

7.4.1 Flow rate

Differential pressure measurements can be made for flow rate determination when a fluid flows through a restriction. The restriction produces an increase in pressure which can be directly related to flow rate. Figure 7.4 shows examples of commonly used restrictions; (*a*) orifice plate, (*b*) Venturi tube, (*c*) flow nozzle, and (*d*) Dall tube.

The orifice plate is normally a simple metal diaphragm with a constricting hole. The diaphragm is normally clamped between pipe flanges to give easy access. The differential pressure ports can be located in the flange on either side of the orifice plate as shown in Fig. 7.4*a*, or alternatively, at specific locations in the pipe on either side of the flange determined by the flow patterns (named vena contracta). A differential pressure gauge is used to measure the difference in pressure between the two ports; the differential pressure gauge can be calibrated in flow rates. The lagging edge of the hole in the diaphragm is beveled to minimize turbulence. In fluids the hole is normally centered in the diaphragm, see Fig. 7.5*a*. However, if the fluid contains particulates, the hole could be placed at the bottom of the pipe to prevent a build up of particulates as in Fig. 7.5*b*. The hole can also be in the form of a semicircle having the same diameter as the pipe and located at the bottom of the pipe as in Fig. 7.5*c*.

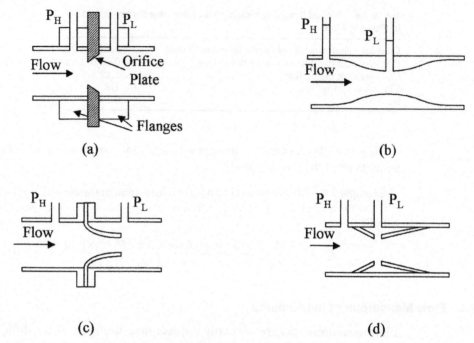

Figure 7.4 Types of constrictions used in flow rate measuring devices (a) orifice plate, (b) Venturi tube, (c) flow nozzle, and (d) Dall tube.

The Venturi tube shown in Fig. 7.4b uses the same differential pressure principle as the orifice plate. The Venturi tube normally uses a specific reduction in tube size, and is not used in larger diameter pipes where it becomes heavy and excessively long. The advantages of the Venturi tube are its ability to handle large amounts of suspended solids, it creates less turbulence and hence less insertion loss than the orifice plate. The differential pressure taps in the Venturi tube are located at the minimum and maximum pipe diameters. The Venturi tube has good accuracy but has a high cost.

The flow nozzle is a good compromise on the cost and accuracy between the orifice plate and the Venturi tube for clean liquids. It is not normally used with

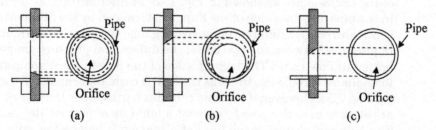

Figure 7.5 Orifice shapes and locations used (a) with fluids and (b) and (c) with suspended solids.

suspended particles. Its main use is the measurement of steam flow. The flow nozzle is shown in Fig. 7.4c.

The Dall tube shown in Fig. 7.4d has the lowest insertion loss but is not suitable for use with slurries.

Typical ratios (beta ratios, which are the diameter of the orifice opening divided by the diameter of the pipe) for the size of the constriction to pipe size in flow measurements are normally between 0.2 and 0.6. The ratios are chosen to give high enough pressure drops for accurate flow measurements but are not high enough to give turbulence. A compromise is made between high beta ratios (d/D) which give low differential pressures and low ratios which give high differential pressures, but can create high losses.

To summarize, the orifice is the simplest, cheapest, easiest to replace, least accurate, more subject to damage and erosion, and has the highest loss. The Venturi tube is more difficult to replace, most expensive, most accurate, has high tolerance to damage and erosion, and the lowest losses of all the three tubes. The flow nozzle is intermediate between the other two and offers a good compromise. The Dall tube has the advantage of having the lowest insertion loss but cannot be used with slurries.

The elbow can be used as a differential flow meter. Figure 7.6a shows the cross section of an elbow. When a fluid is flowing, there is a differential pressure between the inside and outside of the elbow due to the change in direction of the fluid. The pressure difference is proportional to the flow rate of the fluid. The elbow meter is good for handling particulates in solution, with good wear and erosion resistance characteristics but has low sensitivity.

The pilot static tube shown in Fig. 7.6b is an alternative method of measuring the flow rate, but has some disadvantages in measuring flow, in that it really measures the fluid velocity at the nozzle. Because the velocity varies over the cross section of the pipe, the Pilot static tube should be moved across the pipe to establish an average velocity, or the tube should be calibrated for one area. Other disadvantages are that the tube can become clogged with particulates and the differential pressure between the impact and static pressures for low flow rates

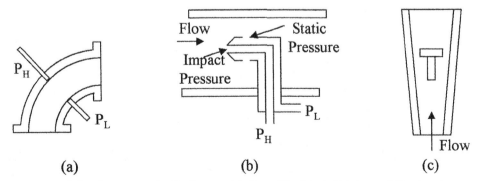

Figure 7.6 Other flow measuring devices are (*a*) elbow, (*b*) pilot static tube, and (*c*) rotameter.

may not be enough to give the required accuracy. The differential pressures in any of the above devices can be measured using the pressure measuring sensors discussed in Chap. 5 (Pressure).

Variable-area meters, such as the rotameter shown in Fig. 7.6c, are often used as a direct visual indicator for flow rate measurements. The rotameter is a vertical tapered tube with a T (or similar) shaped weight. The tube is graduated in flow rate for the characteristics of the gas or liquid flowing up the tube. The velocity of a fluid or gas flowing decreases as it goes higher up the tube, due to the increase in the bore of the tube. Hence, the buoyancy on the weight reduces the higher up the tube it goes. An equilibrium point is eventually reached where the force on the weight due to the flowing fluid is equal to that of the weight, i.e., the higher the flow rate the higher the weight goes up the tube. The position of the weight is also dependent on its size and density, the viscosity and density of the fluid, and the bore and taper of the tube. The Rotameter has a low insertion loss and has a linear relationship to flow rate. In cases where the weight is not visible, i.e., an opaque tube used to reduce corrosion and the like, it can be made of a magnetic material and tracked by a magnetic sensor on the outside of the tube. The rotameter can be used to measure differential pressures across a constriction or flow in both liquids and gases.

An example of rotating flow rate device is the turbine flow meter, which is shown in Fig. 7.7a. The turbine rotor is mounted in the center of the pipe and rotates at a speed proportional to the rate of flow of the fluid or gas passing over the blades. The turbine blades are normally made of a magnetic material or ferrite particles in plastic so that they are unaffected by corrosive liquids. As the blades rotate they can be sensed by a Hall device or magneto resistive element (MRE) sensor attached to the pipe. The turbine should be only used with clean fluids such as gasoline. The rotating flow devices are accurate with good flow operating and temperature ranges, but are more expensive than most of the other devices.

The moving vane is shown in Fig. 7.7b. This device can be used in a pipe configuration as shown or used to measure open channel flow. The vane can be

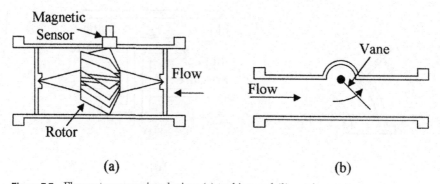

Figure 7.7 Flow rate measuring devices (a) turbine and (b) moving vane.

spring loaded and able to pivot; by measuring the angle of tilt the flow rate can be determined.

Electromagnetic flow meters can only be used in conductive liquids. The device consists of two electrodes mounted in the liquid on opposite sides of the pipe. A magnetic field is generated across the pipe perpendicular to the electrodes as shown in Fig. 7.8a. The conducting fluid flowing through the magnetic field generates a voltage between the electrodes, which can be measured to give the rate of flow. The meter gives an accurate linear output voltage with flow rate. There is no insertion loss and the readings are independent of the fluid characteristics, but it is a relatively expensive instrument.

Vortex flow meters are based on the principle that an obstruction in a fluid or gas flow will cause turbulence or vortices, or in the case of the vortex precession meter (for gases), the obstruction is shaped to give a rotating or swirling motion forming vortices and these can be measured ultrasonically. The frequency of the vortex formation is proportional to the rate of flow and this method is good for high flow rates. At low flow rates the vortex frequency tends to be unstable.

Pressure flow meters use a strain gauge to measure the force on an object placed in a fluid or gas flow. The meter is shown in Fig. 7.8b. The force on the object is proportional to the rate of flow. The meter is low cost with medium accuracy.

7.4.2 Total flow

Includes devices used to measure the total quantity of fluid flowing or the volume of liquid in a flow.

Positive displacement meters use containers of known size, which are filled and emptied for a known number of times in a given time period to give the total flow volume. Two of the more common instruments for measuring total flow are the piston flow meter and the nutating disc flow meter.

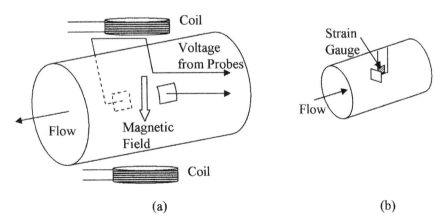

Figure 7.8 Flow measuring devices shown are (a) magnetic flow meter and (b) strain gauge flow meter.

Piston meters consist of a piston in a cylinder. Initially the fluid enters on one side of the piston filling the cylinder, at which point the fluid is diverted to the other side of the piston via valves and the outlet port of the full cylinder is opened. The redirection of fluid reverses the direction of the piston and fills the cylinder on the other side of the piston. The number of times the piston traverses the cylinder in a given time frame gives the total flow. The meter has high accuracy but is expensive.

Nutating disc meters are in the form of a disc that oscillates, allowing a known volume of fluid to pass with each oscillation. The meter is illustrated in Fig. 7.9a. The oscillations can be counted to determine the total volume. This meter can be used to measure slurries but is expensive.

Velocity meters, normally used to measure flow rate, can also be set up to measure the total flow by tracking the velocity and knowing the cross-sectional area of the meter to totalize the flow.

7.4.3 Mass flow

By measuring the flow and knowing the density of a fluid, the mass of the flow can be measured. Mass flow instruments include constant speed impeller turbine wheel-spring combinations that relate the spring force to mass flow and devices that relate heat transfer to mass flow.

Anemometer is an instrument that can be used to measure gas flow rates. One method is to keep the temperature of a heating element in a gas flow constant and measure the power required. The higher the flow rate, the higher the amount of heat required. The alternative method (hot-wire anemometer) is to measure the incident gas temperature and the temperature of the gas down stream from a heating element; the difference in the two temperatures can be related to the flow rate. Micro-machined anemometers are now widely used in automobiles for the measurement of air intake mass. The advantages of this type of sensor are that they are very small, have no moving parts, pose little obstruction to flow, have a low thermal time constant, and are very cost effective along with good longevity.

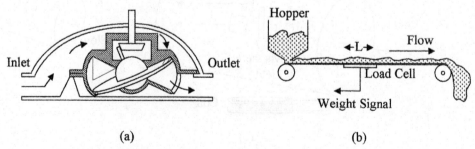

Figure 7.9 Illustrations show (*a*) the cross section of a nutating disc for the measurement of total flow and (*b*) conveyer belt system for the measurement of dry particulate flow rate.

7.4.4 Dry particulate flow rate

Dry particulate flow rate can be measured as the particulate are being carried on a conveyer belt with the use of a load cell. This method is illustrated in Fig. 7.9b. To measure flow rate it is only necessary to measure the weight of material on a fixed length of the conveyer belt.

The flow rate Q is given by

$$Q = \frac{WR}{L} \qquad (7.14)$$

where W = weight of material on length of the weighing platform
L = length of the weighing platform
R = speed of the conveyer belt

Example 7.7 A conveyer belt is traveling at 19 cm/s, a load cell with a length of 1.1 m is reading 3.7 kgm. What is the flow rate of the material on the belt?

$$Q = \frac{3.7 \times 19}{100 \times 1.1} \text{ kg/s} = 0.64 \text{ kg/s}$$

7.4.5 Open channel flow

Open channel flow occurs when the fluid flowing is not contained as in a pipe but is in an open channel. Flow rates can be measured using constrictions as in contained flows. A weir used for open channel flow is shown in Fig. 7.10a. This device is similar in operation to an orifice plate. The flow rate is determined by measuring the differential pressures or liquid levels on either side of the constriction. A Parshall flume is shown in Fig. 7.10b, which is similar in shape to a Venturi tube. A paddle wheel or open flow nozzle are alternative methods of measuring open channel flow rates.

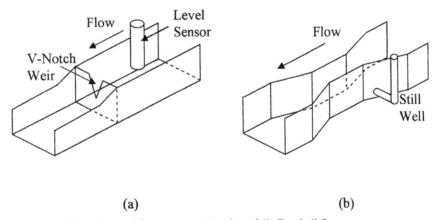

Figure 7.10 Open channel flow sensors (*a*) weir and (*b*) Parshall flume.

7.5 Application Considerations

Many different types of sensors can be used for flow measurements. The choice of any particular device for a specific application depends on a number of factors such as- reliability, cost, accuracy, pressure range, temperature, wear and erosion, energy loss, ease of replacement, particulates, viscosity, and so forth.

7.5.1 Selection

The selection of a flow meter for a specific application to a large extent will depend on the required accuracy and the presence of particulates, although the required accuracy is sometimes down graded because of cost. One of the most accurate meters is the magnetic flow meter which can be accurate to 1 percent of full scale reading or deflection (FSD). The meter is good for low flow rates, with high viscosities and has low energy loss, but is expensive and requires a conductive fluid.

The turbine gives high accuracies and can be used when there is vapor present, but the turbine is better with clean low viscosity fluids. Table 7.5 gives a comparison of flow meter characteristics.

The general purpose and most commonly used devices are the pressure differential sensors used with pipe constrictions. These devices will give accuracies in the 3 percent range when used with solid state pressure sensors which convert the readings directly into electrical units or the rotameter for direct visual reading. The Venturi tube has the highest accuracy and least energy loss followed by the flow nozzle and the orifice plate. For cost effectiveness the devices are in the reverse order. If large amounts of particulates are present, the Venture tube is preferred. The differential pressure devices operate best between 30 and 100 percent of the flow range. The elbow should also be considered in these applications.

Gas flow can be best measured with an anemometer. Solid-state anemometers are now available with good accuracy, are very small in size, and are cost effective.

TABLE 7.5 Summary of Flow Meter Characteristics

Meter type	Range	Accuracy percent	Comments
Orifice plate	3 to 1	±3 FSD	Low cost and accuracy
Venturi tube	3 to 1	±1 FSD	High cost, good accuracy, low losses
Flow nozzle	3 to 1	±2 FSD	Medium cost, accuracy
Dall tube	3 to 1	±2 FSD	Medium cost, accuracy, low losses
Elbow	3 to 1	±6 –10 FSD	Low cost, losses, sensitivity
Pilot static tube	3 to 1	±4 FSD	Low sensitivity
Rotameter	10 to 1	±2 of rate	Low losses, line of sight
Turbine meter	10 to 1	±2 FSD	High accuracy, low losses
Moving vane	5 to 1	±10 FSD	Low cost, low accuracy
Electromagnetic	30 to 1	±0.5 of rate	Conductive fluid, low losses, high cost
Vortex meter	20 to 1	±0.5 of rate	Poor at low flow rates
Strain gauge	3 to 1	±2 FSD	Low cost, accuracy
Nutating disc	5 to 1	±3 FSD	High accuracy, cost
Anemometer	100 to 1	±2 of rate	Low losses, fast response

For open channel applications the flume is the most accurate and best if particulates are present, but is the most expensive.

Particular attention should also be given to manufacturers specifications and application notes.

7.5.2 Installation

Because of the turbulence generated by any type of obstruction in an otherwise smooth pipe, attention has to be given to the placement of flow sensors. The position of the pressure taps can be critical for accurate measurements. The manufacturer's recommendations should be followed during installation. In differential pressure sensing devices the upstream tap should be one to three pipe diameters from the plate or constriction and the down stream tap up to eight pipe diameters from the constriction.

To minimize the pressure fluctuations at the sensor, it is desirable to have a straight run of 10 to15 pipe diameters on either side of the sensing device. It may also be necessary to incorporate laminar flow planes into the pipe to minimize flow disturbances and dampening devices to reduce flow fluctuations to an absolute minimum.

Flow nozzles may require a vertical installation if gases or particulates are present. To allow gases to pass through the nozzle, it should be facing upwards and for particulates, downwards.

7.5.3 Calibration

Flow meters need periodic calibration. This can be done by using another calibrated meter as a reference or by using a known flow rate. Accuracy can vary over the range of the instrument and with temperature and specific weight changes in the fluid, which may all have to be taken into account. Thus, the meter should be calibrated over temperature as well as range, so that the appropriate corrections can be made to the readings. A spot check of the readings should be made periodically to check for instrument drift that may be caused by the instrument going out of calibration, particulate build up, or erosion.

Summary

This chapter discussed the measurement of the flow of fluids in closed and open channels and gases in closed channels. The basic terms, standards, formulas, and laws associated with flow rates are given. Instruments used in the measurement of flow rates are described, as well as considerations in instrument selection for flow measurement are discussed.

The saliant points discussed in this chapter are as follows:

1. The relation of the Reynolds number to physical parameters and its use for determining laminar or turbulent flow in fluids

2. The development of the Bernoulli equation from the concept of the conservation of energy, and modification of the equation to allow for losses in liquid flow

3. Definitions of the terms and standards used in the measurement of the flow of liquids and slurries
4. Difference between flow rates, total flow, and mass flow and the instruments used to measure total flow and mass flow in liquids and gases
5. Various types of flow measuring instruments including the use of restrictions and flow meters for direct and indirect flow measurements
6. Open channel flow and devices used to measure open channel flow rates
7. Comparison of sensor characteristics and considerations in the selection of flow instruments for liquids and slurries and installation precautions

Problems

7.1 The flow rate in a 7-in diameter pipe is 3.2 ft^3/s. What is the average velocity in the pipe?

7.2 A 305 liter/min of water flows through a pipe, what is the diameter of the pipe if the velocity of the water in the pipe is 7.3 m/s?

7.3 A pipe delivers 239 gal of water a minute. If the velocity of the water is 27 ft/s, what is the diameter of the pipe?

7.4 What is the average velocity in a pipe, if the diameter of the pipe is 0.82 cm and the flow rate is 90 cm^3/s?

7.5 Water flows in a pipe of 23-cm diameter with an average velocity of 0.73 m/s, the diameter of the pipe is reduced, and the average velocity of the water increases to 1.66 m/s. What is the diameter of the smaller pipe? What is the flow rate?

7.6 The velocity of oil in a supply line changes from 5.1 to 6.3 ft^3/s when going from a large bore to a smaller bore pipe. If the bore of the smaller pipe is 8.1 in diameter, what is the bore of the larger pipe?

7.7 Water in a 5.5-in diameter pipe has a velocity of 97 gal/s; the pipe splits in two to feed two systems. If after splitting, one pipe is 3.2-in diameter and the other 1.8-in diameter, what is the flow rate from each pipe?

7.8 What is the maximum allowable velocity of a liquid in a 3.2-in diameter pipe to ensure laminar flow? Assume the kinematic viscosity of the liquid is 1.7×10^{-5} ft^2/s.

7.9 A copper sphere is dropped from a building 273 ft tall. What will be its velocity on impact with the ground? Ignore air resistance.

7.10 Three hundred and eighty five gallons of water per minute is flowing through a 4.3-in radius horizontal pipe. If the bore of the pipe is reduced to 2.7-in radius and the pressure in the smaller pipe is 93 psig, what is the pressure in the larger section of the pipe?

7.11 Oil with a specific weight of 53 lb/ft^3 is exiting from a pipe whose center line is 17 ft below the surface of the oil. What is the velocity of the oil from the pipe if there is 1.5 ft head loss in the exit pipe?

7.12 A pump in a fountain pumps 109 gal of water a second through a 6.23-in diameter vertical pipe. How high will the water in the fountain go?

7.13 What is the head loss in a 7-in diameter pipe 118 ft long that has a friction factor of 0.027 if the average velocity of the liquid flowing in the pipe is 17 ft/s?

7.14 What is the radius of a pipe, if the head loss is 1.6 ft when a liquid with a friction factor of 0.033 is flowing with an average velocity of 4.3 ft/s through 73 ft of pipe?

7.15 What is the pressure in a 9.7-in bore horizontal pipe, if the bore of the pipe narrows to 4.1 in downstream where the pressure is 65 psig and 28,200 gal of fluid per hour with a specific gravity (SG) of 0.87 is flowing? Neglect losses.

7.16 Fluid is flowing through the following 1-in fittings; 3 threaded ells, 6 tees, 7 globe valves, and 9 unions. If the head loss is 7.2 ft, what is the velocity of the liquid?

7.17 The drag coefficient on a 6.3-in diameter sphere is 0.35. What is its velocity through a liquid with a SG = 0.79 if the drag force is 4.8 lb?

7.18 A square disc is placed in a moving liquid, the drag force on the disc is 6.3 lbs when the liquid has a velocity of 3.4 ft/s. If the liquid has a density of 78.3 lb/ft^3 and the drag coefficient is 0.41, what is the size of the square?

7.19 The 8 × 32-in^3 cylinders in a positive displacement meter assembly are rotating at a rate of 570 revolutions an hour. What is the average flow rate per min?

7.20 Alcohol flows in a horizontal pipe 3.2-in diameter; the diameter of the pipe is reduced to 1.8 in. If the differential pressure between the two sections is 1.28 psi, what is the flow rate through the pipe? Neglect losses.

Chapter 8

Temperature and Heat

Chapter Objectives

This chapter will help you understand the difference between temperature and heat, the units used for their measurement, thermal time constants, and the most common methods used to measure temperature and heat and their standards.

Topics covered in this chapter are as follows:

- The difference between temperature and heat
- The various temperature scales
- Temperature and heat formulas
- The various mechanisms of heat transfer
- Specific heat and heat energy
- Coefficients of linear and volumetric expansion
- The wide variety of temperature measuring devices
- Introduction to thermal time constants

8.1 Introduction

Similar to our every day needs of temperature control for comfort, almost all industrial processes need accurately controlled temperatures. Physical parameters and chemical reactions are temperature dependent, and therefore temperature control is of major importance. Temperature is without doubt the most measured variable, and for accurate temperature control its precise measurement is required. This chapter discusses the various temperature scales used, their relation to each other, methods of measuring temperature, and the relationship between temperature and heat.

8.2 Basic Terms

8.2.1 Temperature definitions

Temperature is a measure of the thermal energy in a body, which is the relative hotness or coldness of a medium and is normally measured in degrees using one of the following scales; Fahrenheit (F), Celsius or Centigrade (C), Rankine (R), or Kelvin (K).

Absolute zero is the temperature at which all molecular motion ceases or the energy of the molecule is zero.

Fahrenheit scale was the first temperature scale to gain acceptance. It was proposed in the early 1700s by Fahrenheit (Dutch). The two points of reference chosen for 0 and 100° were the freezing point of a concentrated salt solution (at sea level) and the internal temperature of oxen (which was found to be very consistent between animals). This eventually led to the acceptance of 32° and 212° (180° range) as the freezing and boiling point, respectively of pure water at 1 atm (14.7 psi or 101.36 kPa) for the Fahrenheit scale. The temperature of the freezing point and boiling point of water changes with pressure.

Celsius or centigrade scale (C) was proposed in mid 1700s by Celsius (Sweden), who proposed the temperature readings of 0° and 100° (giving a 100° scale) for the freezing and boiling points of pure water at 1 atm.

Rankine scale (R) was proposed in the mid 1800s by Rankine. It is a temperature scale referenced to absolute zero that was based on the Fahrenheit scale, i.e., a change of 1°F = a change of 1°R. The freezing and boiling point of pure water are 491.6°R and 671.6°R, respectively at 1 atm, see Fig. 8.1.

Kelvin scale (K) named after Lord Kelvin was proposed in the late 1800s. It is referenced to absolute zero but based on the Celsius scale, i.e., a change of 1°C = a change of 1 K. The freezing and boiling point of pure water are 273.15 K and

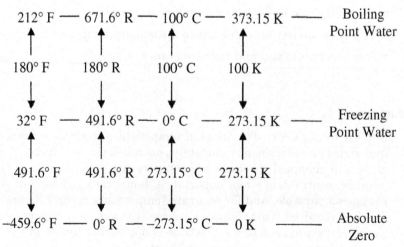

Figure 8.1 Comparison of temperature scales.

373.15 K, respectively, at 1 atm, see Fig. 8.1. The degree symbol can be dropped when using the Kelvin scale.

8.2.2 Heat definitions

Heat is a form of energy; as energy is supplied to a system the vibration amplitude of its molecules and its temperature increases. The temperature increase is directly proportional to the heat energy in the system.

A *British Thermal Unit* (BTU or Btu) is defined as the amount of energy required to raise the temperature of 1 lb of pure water by 1°F at 68°F and at atmospheric pressure. It is the most widely used unit for the measurement of heat energy.

A *calorie unit* (SI) is defined as the amount of energy required to raise the temperature of 1 gm of pure water by 1°C at 4°C and at atmospheric pressure. It is also a widely used unit for the measurement of heat energy.

Joules (SI) are also used to define heat energy and is often used in preference to the calorie, where 1 J (Joule) = 1 W (Watt) × s. This is given in Table 8.1 that gives a list of energy equivalents.

Phase change is the transition of matter from the solid to the liquid to the gaseous states; matter can exist in any of these three states. However, for matter to make the transition from one state up to the next, i.e., solid to liquid to gas, it has to be supplied with energy, or energy removed if the matter is going from gas to liquid to solid. For example, if heat is supplied at a constant rate to ice at 32°F, the ice will start to melt or turn to liquid, but the temperature of the ice liquid mixture will not change until all the ice has melted. Then as more heat is supplied, the temperature will start to rise until the boiling point of the water is reached. The water will turn to steam as more heat is applied but the temperature of the water and steam will remain at the boiling point until all the water has turned to steam, then the temperature of the steam will start to rise above the boiling point. This is illustrated in Fig. 8.2, where the temperature of a substance is plotted against heat input. Material can also change its volume during the change of phase. Some materials bypass the liquid stage and transform directly from solid to gas or gas to solid, this transition is called *Sublimation*.

In a solid, the atoms can vibrate but are strongly bonded to each other so that the atoms or molecules are unable to move from their relative positions. As the temperature is increased, more energy is given to the molecules and their vibration amplitude increases to a point where it can overcome the bonds between the molecules and they can move relative to each other. When this point

TABLE 8.1   Conversion Related to Heat Energy

1 BTU = 252 cal	1 cal = 0.0039 BTU
1 BTU = 1055 J	1 J = 0.000948 BTU
1 BTU = 778 ft-lb	1 ft-lb = 0.001285 BTU
1 cal = 4.19 J	1 J = 0.239 cal
1 ft-lb = 0.324 cal	1 J = 0.738 ft-lb
1 ft-lb = 1.355 J	1 W = 1 J/s

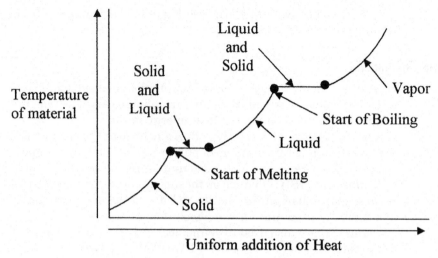

Figure 8.2 Showing the relation between temperature and heat energy.

is reached the material becomes a liquid. The speed at which the molecules move about in the liquid is a measure of their thermal energy. As more energy is imparted to the molecules their velocity in the liquid increases to a point where they can escape the bonding or attraction forces of other molecules in the material and the gaseous state or boiling point is reached.

Specific heat is the quantity of heat energy required to raise the temperature of a given weight of a material by 1°. The most common units are BTUs in the English system, i.e., 1 BTU is the heat required to raise 1 lb of material by 1°F and in the SI system, the calorie is the heat required to raise 1 g of material by 1°C. Thus, if a material has a specific heat of 0.7 cal/g °C, it would require 0.7 cal to raise the temperature of a gram of the material by 1°C or 2.93 J to raise the temperature of the material by 1 k. Table 8.2 gives the specific heat of some common materials; the units are the same in either system.

Thermal conductivity is the flow or transfer of heat from a high temperature region to a low temperature region. There are three basic methods of heat transfer; conduction, convection, and radiation. Although these modes of transfer can be considered separately, in practice two or more of them can be present simultaneously.

TABLE 8.2 Specific Heats of Some Common Materials

Material	Specific heat	Material	Specific heat	Material	Specific heat
Alcohol	0.58–0.6	Aluminum	0.214	Brass	0.089
Glass	0.12–0.16	Cast iron	0.119	Copper	0.092
Gold	0.0316	Lead	0.031	Mercury	0.033
Platinum	0.032	Quartz	0.188	Silver	0.056
Steel	0.107	Tin	0.054	Water	1.0

The units are BTU/lb °F or Calories/g °C.

TABLE 8.3 Thermal Conductivity BTU/h ft °F (W/mK)

Material	Conductivity	Material	Conductivity
Air	0.016 (room temp.) (0.028)	Aluminum	119 (206)
Concrete	0.8 (1.4)	Copper	220 (381)
Water	0.36 (room temp.) (0.62)	Mercury	4.8 (8.3)
Brick	0.4 (0.7)	Steel	26 (45)
Brass	52 (90)	Silver	242 (419)

Conduction is the flow of heat through a material. The molecular vibration amplitude or energy is transferred from one molecule in a material to the next. Hence, if one end of a material is at an elevated temperature, heat is conducted to the cooler end. The thermal conductivity of a material k is a measure of its efficiency in transferring heat. The units can be in BTUs per hour per ft per °F or watts per meter-Kelvin (W/m K) (1 BTU/ft h °F = 1.73 W/mK). Table 8.3 gives typical thermal conductivities for some common materials.

Convection is the transfer of heat due to motion of elevated temperature particles in a material (liquid and gases). Typical examples are air conditioning systems, hot water heating systems, and so forth. If the motion is solely due to the lower density of the elevated temperature material, the transfer is called free or natural convection. If the material is moved by blowers or pumps the transfer is called forced convection.

Radiation is the emission of energy by electromagnetic waves that travel at the speed of light through most materials that do not conduct electricity. For instance, radiant heat can be felt some distance from a furnace where there is no conduction or convection.

8.2.3 Thermal expansion definitions

Linear thermal expansion is the change in dimensions of a material due to temperature changes. The change in dimensions of a material is due to its coefficient of thermal expansion that is expressed as the change in linear dimension (α) per degree temperature change.

Volume thermal expansion is the change in the volume (β) per degree temperature change due to the linear coefficient of expansion. The thermal expansion coefficients for some common materials per degree fahrenheit are given in Table 8.4. The coefficients can also be expressed as per degree Celsius.

TABLE 8.4 Thermal Coefficients of Expansion per Degree Fahrenheit

Material	Linear ($\times 10^{-6}$)	Volume ($\times 10^{-6}$)	Material	Linear ($\times 10^{-6}$)	Volume ($\times 10^{-6}$)
Alcohol	—	61–66	Aluminum	12.8	—
Brass	10	—	Cast iron	5.6	20
Copper	9.4	29	Glass	5	14
Gold	7.8	—	Lead	16	—
Mercury	—	100	Platinum	5	15
Quartz	0.22	—	Silver	11	32
Steel	6.1	—	Tin	15	38

8.3 Temperature and Heat Formulas

8.3.1 Temperature

The need to convert from one temperature scale to another is a common everyday occurrence. The conversion factors are as follows:

To convert °F to °C

$$°C = (°F - 32)5/9 \tag{8.1}$$

To convert °C to °F

$$°F = (°C \times 9/5) + 32 \tag{8.2}$$

To convert °F to °R

$$°R = °F + 459.6 \tag{8.3}$$

To convert °C to K

$$K = °C + 273.15 \tag{8.4}$$

To convert K to °R

$$°R = 1.8 \times K \tag{8.5}$$

To convert °R to K

$$K = 0.555 \times °R \tag{8.6}$$

Example 8.1 What temperature in Kelvin corresponds to 115°F?
From Eq.(8.1)

$$°C = (115 - 32)5/9 = 46.1°C$$

From Eq.(8.4)

$$K = 46.1 + 273.15 = 319.25 \text{ K}$$

8.3.2 Heat transfer

The amount of heat needed to raise or lower the temperature of a given weight of a body can be calculated from the following equation:

$$Q = WC(T_2 - T_1) \tag{8.7}$$

where W = weight of the material
C = specific heat of the material
T_2 = final temperature of the material
T_1 = initial temperature of the material

Example 8.2 What is the heat required to raise the temperature of a 1.5 kg mass 120°C if the specific heat of the mass is 0.37 cal/g°C?

$$Q = 1.5 \times 1000 \text{ g} \times 0.37 \text{ cal/g°C} \times 120°C = 66,600 \text{ cal}$$

As always, care must be taken in selecting the correct units. Negative answers indicate extraction of heat or heat loss.

Heat conduction through a material is derived from the following relationship:

$$Q = \frac{-kA(T_2 - T_1)}{L} \tag{8.8}$$

where Q = rate of heat transfer
k = thermal conductivity of the material
A = cross-sectional area of the heat flow
T_2 = temperature of the material distant from the heat source
T_1 = temperature of the material adjacent to heat source
L = length of the path through the material

Note; the negative sign in the Eq. (8.8) indicates a positive heat flow.

Example 8.3 A furnace wall 12 ft^2 in area and 6-in thick has a thermal conductivity of 0.14 BTU/h ft°F. What is the heat loss if the furnace temperature is 1100°F and the outside of the wall is 102°F?

$$Q = \frac{-kA(T_2 - T_1)}{L}$$

$$Q = \frac{-0.14 \times 12(102 - 1100)}{0.5} = 3{,}353.3 \text{ BTU/h}$$

Example 8.4 The outside wall of a room is 4 × 3 m and 0.35 m thick. What is the energy loss per hour if the inside and outside temperatures are 35°C and −40°C respectively? Assume the conductivity of the wall is 0.13 W/mK.

$$Q = \frac{-kA(T_2 - T_1)}{L}$$

$$Q = \frac{-0.13 \text{ W/mK} \times 4 \text{ m} \times 3 \text{ m} \times (-40 - 35) \text{ K}}{0.35 \text{ m}} \times \frac{60 \times 60 \text{ J/s}}{\text{W} \times \text{h}}$$

$$Q = 1203 \text{ kJ/h}$$

Heat convection calculations in practice are not as straight forward as conduction. However, heat convection is given by

$$Q = hA \, (T_2 - T_1) \tag{8.9}$$

where Q = convection heat transfer rate
h = coefficient of heat transfer
h = heat transfer area
$T_2 - T_1$ = temperature difference between the source and final temperature of the flowing medium

It should be noted that in practice the proper choice for h is difficult because of its dependence on a large number of variables (such as density, viscosity, and specific heat). Charts are available for h. However, experience is needed in their application.

Example 8.5 How much heat is transferred from a 25 ft × 24 ft surface by convection if the temperature difference between the front and back surfaces is 40°F and the surface has a heat transfer rate of 0.22 BTU/h ft²°F?

$$Q = 0.22 \times 25 \times 24 \times 40 = 39{,}600 \text{ BTU/h}$$

Heat radiation depends on surface color, texture, shapes involved and the like. Hence, more information than the basic relationship for the transfer of radiant heat energy given below should be factored in. The radiant heat transfer is given by

$$Q = CA\left(T_2^4 - T_1^4\right) \tag{8.10}$$

where Q = heat transferred
C = radiation constant (depends on surface color, texture, units used, and the like)
A = area of the radiating surface
T_2 = absolute temperature of the radiating surface
T_1 = absolute temperature of the receiving surface

Example 8.6 The radiation constant for a furnace is 0.23×10^{-8} BTU/h ft²°F⁴, the radiating surface area is 25 ft². If the radiating surface temperature is 750°F and the room temperature is 75°F, how much heat is radiated?

$$Q = 0.23 \times 10^{-8} \times 25[\{750 + 460\}^4 - \{75 + 460\}^4]$$
$$Q = 5.75 \times 10^{-8} [222 \times 10^{10} - 8.4 \times 10^{10}] = 1.2 \times 10^5 \text{ BTU/h}$$

Example 8.7 What is the radiation constant for a wall 5 m × 4 m, if the radiated heat loss is 62.3 MJ/h when the wall and ambient temperatures are 72°C and 5°C?

$$62.3 \text{ MJ/h} = 17.3 \text{ kW} = C \times 20 \: [\{72 + 273.15\}^4 - \{5 + 273.15\}^4]$$
$$C = 17.3 \times 10^3/20 \: (1.419 \times 10^{10} - 0.598 \times 10^{10})$$
$$C = 17.3/16.41 \times 10^7 = 1.05 \times 10^{-7} \text{ W/m}^2 \text{ K}^4$$

8.3.3 Thermal expansion

Linear expansion of a material is the change in linear dimension due to temperature changes and can be calculated from the following formula:

$$L_2 = L_1 \: [1 + \alpha \: (T_2 - T_1)] \tag{8.11}$$

where L_2 = final length
L_1 = initial length
α = coefficient of linear thermal expansion
T_2 = final temperature
T_1 = initial temperature

Volume expansion in a material due to changes in temperature is given by

$$V_2 = V_1 \: [1 + \beta \: (T_2 - T_1)] \tag{8.12}$$

where V_2 = final volume
V_1 = initial volume
β = coefficient of volumetric thermal expansion
T_2 = final temperature
T_1 = initial temperature

Example 8.8 Calculate the length and volume for a 200 cm on a side copper cube at 20°C, if the temperature is increased to 150°C.

$$\text{New length} = 200(1 + 9.4 \times 10^{-6} \times [150 - 20] \times 9/5)$$
$$= 200(1 + .0022) = 200.44 \text{ cm}$$
$$\text{New volume} = 200^3(1 + 29 \times 10^{-6} \times [150 - 20] \times 9/5)$$
$$= 200^3(1 + .0068) = 8054400 \text{ cm}^3$$

In a gas, the relation between the pressure, volume, and temperature of the gas is given by

$$\frac{P_1 V_1}{T_1} = \frac{P_2 V_2}{T_2} \qquad (8.13)$$

where P_1 = initial pressure
V_1 = initial volume
T_1 = initial absolute temperature
P_2 = final pressure
V_2 = final volume
T_2 = final absolute temperature

8.4 Temperature Measuring Devices

There are several methods of measuring temperature that can be categorized as follows:

1. Expansion of a material to give visual indication, pressure, or dimensional change
2. Electrical resistance change
3. Semiconductor characteristic change
4. Voltage generated by dissimilar metals
5. Radiated energy

Thermometer is often used as a general term given to devices for measuring temperature. Examples of temperature measuring devices are described below.

8.4.1 Thermometers

Mercury in glass was by far the most common direct visual reading thermometer (if not the only one). The device consisted of a small bore graduated glass tube with a small bulb containing a reservoir of mercury. The coefficient of expansion of mercury is several times greater than the coefficient of expansion of glass, so that as the temperature increases the mercury rises up the tube giving a relatively low cost and accurate method of measuring temperature. Mercury also has the advantage of not wetting the glass, and hence, cleanly traverses the glass tube without breaking into globules or coating the tube. The operating range of the mercury thermometer is from −30 to 800°F (−35 to 450°C) (freezing point of mercury −38°F [−38°C]). The toxicity of mercury, ease of breakage, the introduction of cost effective, accurate, and easily read digital thermometers has brought about the demise of the mercury thermometer.

Liquids in glass devices operate on the same principle as the mercury thermometer. The liquids used have similar properties to mercury, i.e., high linear coefficient of expansion, clearly visible, nonwetting, but are nontoxic. The liquid in glass thermometers is used to replace the mercury thermometer and to extend its operating range. These thermometers are accurate and with different liquids (each type of liquid has a limited operating range) can have an operating range of from −300 to 600°F (−170 to 330°C).

Bimetallic strip is a type of temperature measuring device that is relatively inaccurate, slow to respond, not normally used in analog applications to give remote indication, and has hysteresis. The bimetallic strip is extensively used in ON/OFF applications not requiring high accuracy, as it is rugged and cost effective. These devices operate on the principle that metals are pliable and different metals have different coefficients of expansion (see Table 8.4). If two strips of dissimilar metals such as brass and invar (copper-nickel alloy) are joined together along their length, they will flex to form an arc as the temperature changes; this is shown in Fig. 8.3a. Bimetallic strips are usually configured as a spiral or helix for compactness and can then be used with a pointer to make a cheap compact rugged thermometer as shown in Fig. 8.3b. Their operating range is from −180 to 430°C and can be used in applications from oven thermometers to home and industrial control thermostats.

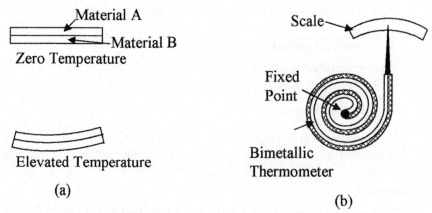

Figure 8.3 Shows (a) the effect of temperature change on a bimetallic strip and (b) bimetallic strip thermometer.

8.4.2 Pressure-spring thermometers

These thermometers are used where remote indication is required, as opposed to glass and bimetallic devices which give readings at the point of detection. The pressure-spring device has a metal bulb made with a low coefficient of expansion material with a long metal tube, both contain material with a high coefficient of expansion; the bulb is at the monitoring point. The metal tube is terminated with a spiral Bourdon tube pressure gage (scale in degrees) as shown in Fig. 8.4a. The pressure system can be used to drive a chart recorder, actuator, or a potentiometer wiper to obtain an electrical signal. As the temperature in the bulb increases, the pressure in the system rises, the pressure rise being proportional to the temperature change. The change in pressure is sensed by the Bourdon tube and converted to a temperature scale. These devices can be accurate to 0.5 percent and can be used for remote indication up to 100 m but must be calibrated, as the stem and Bourdon tube are temperature sensitive.

There are three types or classes of pressure-spring devices. These are as follows:

Class 1   Liquid filled
Class 2   Vapor pressure
Class 3   Gas filled

Liquid filled thermometer works on the same principle as the liquid in glass thermometer, but is used to drive a Bourdon tube. The device has good linearity and accuracy and can be used up to 550°C.

Vapor-pressure thermometer system is partially filled with liquid and vapor such as methyl chloride, ethyl alcohol, ether, toluene, and so on. In this system the lowest operating temperature must be above the boiling point of the liquid

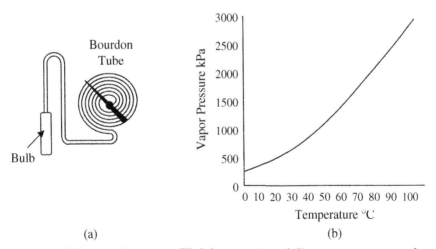

Figure 8.4 Illustrates (a) pressure filled thermometer and (b) vapor pressure curve for methyl chloride.

and the maximum temperature is limited by the critical temperature of the liquid. The response time of the system is slow, being of the order of 20 s. The temperature pressure characteristic of the thermometer is nonlinear as shown in the vapor pressure curve for methyl chloride in Fig. 8.4b.

Gas thermometer is filled with a gas such as nitrogen at a pressure range of 1000 to 3350 kPa at room temperature. The device obeys the basic gas laws for a constant volume system [Eq.(8.15), $V_1 = V_2$] giving a linear relationship between absolute temperature and pressure.

8.4.3 Resistance temperature devices

Resistance temperature devices (RTD) are either a metal film deposited on a former or are wire-wound resistors. The devices are then sealed in a glass-ceramic composite material. The electrical resistance of pure metals is positive, increasing linearly with temperature. Table 8.5 gives the temperature coefficient of resistance of some common metals used in resistance thermometers. These devices are accurate and can be used to measure temperatures from –300 to 1400°F (–170 to 780°C).

In a resistance thermometer the variation of resistance with temperature is given by

$$R_{T2} = R_{T1}(1 + \text{Coeff.} [T_2 - T_1]) \qquad (8.14)$$

where R_{T2} is the resistance at temperature T_2 and R_{T1} is the resistance at temperature T_1.

Example 8.9 What is the resistance of a platinum resistor at 250°C, if its resistance at 20°C is 1050 Ω?

$$\begin{aligned}\text{Resistance at } 250°C &= 1050(1 + 0.00385\,[250 - 20]) \\ &= 1050(1 + 0.8855) \\ &= 1979.775 \text{ Ω}\end{aligned}$$

Resistance devices are normally measured using a Wheatstone bridge type of system, but are supplied from a constant current source. Care should also be taken to prevent electrical current from heating the device and causing erroneous readings. One method of overcoming this problem is to use a pulse technique. When using this method the current is turned ON for say 10 ms every 10 s, and the sensor resistance is measured during this 10 ms time period. This reduces the internal heating effects by 1000 to 1 or the internal heating error by this factor.

TABLE 8.5 Temperature Coefficient of Resistance of Some Common Metals

Material	Coeff. per degree Celsius	Material	Coeff. per degree Celsius
Iron	0.006	Tungsten	0.0045
Nickel	0.005	Platinum	0.00385

8.4.4 Thermistors

Thermistors are a class of metal oxide (semiconductor material) which typically have a high negative temperature coefficient of resistance, but can also be positive. Thermistors have high sensitivity which can be up to 10 percent change per degree Celsius, making them the most sensitive temperature elements available, but with very nonlinear characteristics. The typical response times is 0.5 to 5 s with an operating range from −50 to typically 300°C. Devices are available with the temperature range extended to 500°C. Thermistors are low cost and manufactured in a wide range of shapes, sizes, and values. When in use care has to be taken to minimize the effects of internal heating. Thermistor materials have a temperature coefficient of resistance (α) given by

$$\alpha = \frac{\Delta R}{R_S}\left(\frac{1}{\Delta T}\right) \tag{8.15}$$

where ΔR is the change in resistance due to a temperature change ΔT and R_S the material resistance at the reference temperature.

The nonlinear characteristics are as shown in Fig. 8.5 and make the device difficult to use as an accurate measuring device without compensation, but its sensitivity and low cost makes it useful in many applications. The device is normally used in a bridge circuit and padded with a resistor to reduce its nonlinearity.

8.4.5 Thermocouples

Thermocouples are formed when two dissimilar metals are joined together to form a junction. An electrical circuit is completed by joining the other ends of the dissimilar metals together to form a second junction. A current will flow in the circuit if the two junctions are at different temperatures as shown in Fig. 8.6a.

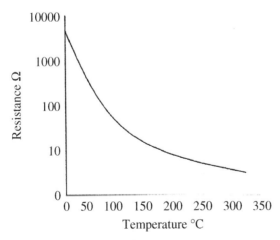

Figure 8.5 Thermistor resistance temperature curve.

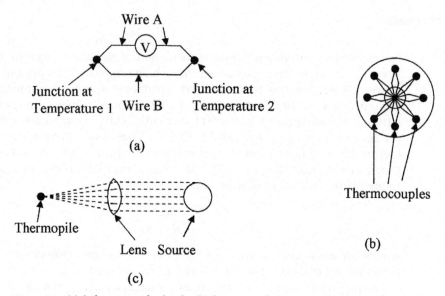

Figure 8.6 (a) A thermocouple circuit, (b) thermocouples connected to form a thermopile, and (c) focusing EM rays onto a thermopile.

The current flowing is the result of the difference in electromotive force developed at the two junctions due to their temperature difference. In practice, the voltage difference between the two junctions is measured; the difference in the voltage is proportional to the temperature difference between the two junctions. Note that the thermocouple can only be used to measure temperature differences. However, if one junction is held at a reference temperature the voltage between the thermocouples gives a measurement of the temperature of the second junction.

Three effects are associated with thermocouples. They are as follows:

1. *Seebeck effect.* It states that the voltage produced in a thermocouple is proportional to the temperature between the two junctions.

2. *Peltier effect.* It states that if a current flows through a thermocouple one junction is heated (puts out energy) and the other junction is cooled (absorbs energy).

3. *Thompson effect.* It states that when a current flows in a conductor along which there is a temperature difference, heat is produced or absorbed, depending upon the direction of the current and the variation of temperature.

In practice, the Seebeck voltage is the sum of the electromotive forces generated by the Peltier and Thompson effects. There are a number of laws to be observed in thermocouple circuits. Firstly, the law of intermediate temperatures states that the thermoelectric effect depends only on the temperatures of the junctions and is not affected by the temperatures along the leads. Secondly, the law of intermediate metals states that metals other than those making up the thermocouples can be used in the circuit as long as their junctions are at the same temperature, i.e., other types of metals can be used for interconnections

TABLE 8.6 Operating Ranges for Thermocouples and Seebeck Coefficients

Type	Approx. range (°C)	Seebeck coefficient (μV/°C)
Copper–Constantan (T)	−140 to 400	40 (−59 to 93) ±1°C
Chromel–Constantan (E)	−180 to 1000	62 (0 to 360) ±2°C
Iron–Constantan (J)	30 to 900	51 (0 to 277) ±2°C
Chromel–Alumel (K)	30 to 1400	40 (0 to 277) ±2°C
Nicrosil–Nisil (N)	30 to 1400	38 (0 to 277) ±2°C
Platinum (rhodium 10%)–Platinum (S)	30 to 1700	7 (0 to 538) ±3°C
Platinum (rhodium 13%)–Platinum (R)	30 to 1700	7 (0 to 538) ±3°C

and tag strips can be used without adversely affecting the output voltage from the thermocouple. The various types of thermocouples are designated by letters. Tables of the differential output voltages for different types of thermocouples are available from manufacturer's thermocouple data sheets. Table 8.6 lists some thermocouple materials and their Seebeck coefficient. The operating range of the thermocouple is reduced to the figures in brackets if the given accuracy is required. For operation over the full temperature range the accuracy would be reduced to about ±10 percent without linearization.

Thermopile is a number of thermocouples connected in series, to increase the sensitivity and accuracy by increasing the output voltage when measuring low temperature differences. Each of the reference junctions in the thermopile is returned to a common reference temperature as shown in Fig. 8.6b.

Radiation can be used to sense temperature. The devices used are pyrometers using thermocouples or color comparison devices.

Pyrometers are devices that measure temperature by sensing the heat radiated from a hot body through a fixed lens that focuses the heat energy on to a thermopile; this is a noncontact device. Furnace temperatures, for instance, are normally measured through a small hole in the furnace wall. The distance from the source to the pyrometer can be fixed and the radiation should fill the field of view of the sensor. Figure 8.6c shows the focusing lens and thermocouple set up in a thermopile.

Figure 8.7 shows plots of the electromotive force (emf) versus temperature of some of the types of thermocouples available.

8.4.6 Semiconductors

Semiconductors have a number of parameters that vary linearly with temperature. Normally the reference voltage of a zener diode or the junction voltage variations are used for temperature sensing. Semiconductor temperature sensors have a limited operating range from −50 to 150°C but are very linear with accuracies of ±1°C or better. Other advantages are that electronics can be integrated onto the same die as the sensor giving high sensitivity, easy interfacing to control systems, and making different digital output configurations possible. The thermal time constant varies from 1 to 5 s, internal dissipation can also cause up to 0.5°C offset. Semiconductor devices are also rugged with good longevity and are inexpensive. For the above reasons the semiconductor sensor is used extensively in many applications including the replacement of the mercury in glass thermometer.

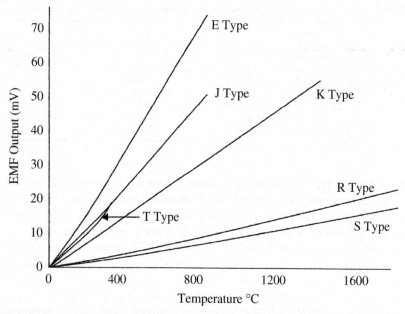

Figure 8.7 Thermocouple emf versus temperature for various types.

8.5 Application Considerations

8.5.1 Selection

In process control a wide selection of temperature sensors are available. However, the required range, linearity, and accuracy can limit the selection. In the final selection of a sensor, other factors may have to be taken into consideration, such as remote indication, error correction, calibration, vibration sensitivity, size, response time, longevity, maintenance requirements, and cost. The choice of sensor devices in instrumentation should not be degraded from a cost standpoint. Process control is only as good as the monitoring elements.

8.5.2 Range and accuracy

Table 8.7 gives the temperature ranges and accuracies of temperature sensors. The accuracies shown are with minimal calibration or error correction. The ranges in some cases can be extended with the use of new materials. Table 8.8 gives a summary of temperature sensor characteristics.

8.5.3 Thermal time constant

A temperature detector does not react immediately to a change in temperature. The reaction time of the sensor or thermal time constant is a measure of the time it takes for the sensor to stabilize internally to the external temperature change, and is determined by the thermal mass and thermal conduction resistance of the device. Thermometer bulb size, probe size, or protection well can affect the

TABLE 8.7 Temperature Range and Accuracy of Temperature Sensors

Sensor type		Range (degree Celsius)	Accuracy (FSD)
Expansion	Mercury in glass	−35 to 430	±1%
	Liquid in glass	−180 to 500	±1%
	Bimetallic	−180 to 600	±20%
Pressure–spring	Liquid filled	−180 to 550	±0.5%
	Vapor pressure	−180 to 550	±2.0%
	Gas filled	−180 to 550	±0.5%
Resistance	Metal resistors	−200 to 800	±5%
	Platinum	−180 to 650	±0.5%
	Nickel	−180 to 320	±1%
	Copper	−180 to 320	±0.2%
Thermistor		0 to 500	±25%
Thermocouple		−60 to 540	±1%
		−180 to 2500	±10%
Semiconductor IC		−40 to 150	±1%

response time of the reading, i.e., a large bulb contains more liquid for better sensitivity, but this will also increase the time constant taking longer to fully respond to a temperature change.

The thermal time constant is related to the thermal parameters by the following equation:

$$t_c = \frac{mc}{kA} \tag{8.16}$$

where t_c = thermal time constant
m = mass
c = specific heat
k = heat transfer coefficient
A = area of thermal contact

TABLE 8.8 Summary of Sensor Characteristics

Type	Linearity	Advantages	Disadvantages
Bimetalic	Good	Low cost, rugged, and wide range	Local measurement or for ON/OFF switching only
Pressure	Medium	Accurate and wide range	Needs temperature compensation and vapor is nonlinear
Resistance	Very good	Stable, wide range, and accurate	Slow response, low sensitivity, expensive, self heating, and limited range
Thermistor	Poor	Low cost, small, high sensitivity, and fast response	Nonlinear, range, and self heating
Thermocouple	Good	Low cost, rugged, and very wide range	Low sensitivity and reference needed
Semiconductor	Excellent	Low cost, sensitive and easy to interface	Self heating, Slow response, range, and power source

When the temperature changes rapidly, the temperature output reading of a thermal sensor is given by

$$T - T_2 = (T_1 - T_2)e^{-t/tc} \qquad (8.17)$$

where T = temperature reading
T_1 = initial temperature
T_2 = true system temperature
t = time from when the change occurred

The time constant of a system t_c is considered as the time it takes for the system to reach 63.2 percent of its final temperature value after a temperature change, i.e., a copper block is held in an ice–water bath until its temperature has stabilized at 0°C, it is then removed and placed in a 100°C steam bath, the temperature of the copper block will not immediately go to 100°C, but its temperature will rise on an exponential curve as it absorbs energy from the steam, until after some time period (its time constant) it will reach 63.2°C, aiming to eventually reach 100°C. This is shown in the graph (line A) in Fig. 8.8. During the second time constant the copper will rise another 63.2 percent of the remaining temperature to get to equilibrium, i.e., (100 − 63.2) 63.2 percent = 23.3°C, or at the end of 2 time constant periods, the temperature of the copper will be 86.5°C. At the end of 3 periods the temperature will be 95° and so on. Also shown in Fig. 8.8 is a second line B for the copper, the time constants are the same but the final aiming temperature is 50°C. The time to stabilize is the same in both cases. Where a fast response time is required, thermal time constants can be a serious problem as in some cases they can be of several seconds duration. Correction may have to be applied to the output reading electronically to correct for the thermal time constant to obtain a faster response. This can be

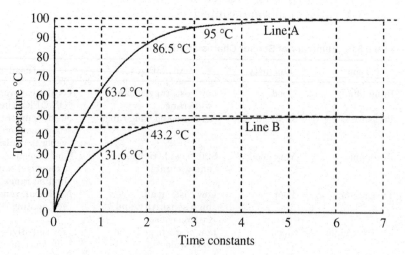

Figure 8.8 Shows the response time to changes in temperature.

done by measuring the rate of rise of the temperature indicated by the sensor and extrapolating the actual aiming temperature.

The thermal time constant of a body is similar to an electrical time constant which is discussed in the chapter on electricity under electrical time constants.

8.5.4 Installation

Care must be taken in locating the sensing portion of the temperature sensor, it should be fully encompassed by the medium whose temperature is being measured, and not be in contact with the walls of the container. The sensor should be screened from reflected heat and radiant heat if necessary. The sensor should also be placed downstream from the fluids being mixed, to ensure that the temperature has stabilized, but as close as possible to the point of mixing, to give as fast as possible temperature measurement for good control. A low thermal time constant in the sensor is necessary for a quick response.

Compensation and calibration may be necessary when using pressure-spring devices with long tubes especially when accurate readings are required.

8.5.5 Calibration

Temperature calibration can be performed on most temperature sensing devices by immersing them in known temperature standards which are the equilibrium points of solid/liquid or liquid/gas mixtures, which is also known as the triple point. Some of these are given in Table 8.9. Most temperature sensing devices are rugged and reliable, but can go out of calibration due to leakage during use or contamination during manufacture and should therefore be checked on a regular basis.

8.5.6 Protection

In some applications, temperature sensing devices are placed in wells or enclosures to prevent mechanical damage or for ease of replacement. This kind of protection can greatly increase the system response time, which in some circumstances may be unacceptable. Sensors may need also to be protected from over temperature, so that a second more rugged device may be needed to protect the main sensing device. Semiconductor devices may have built in over

TABLE 8.9 Temperature Scale Calibration Points

Calibration material	Temperature			
	K	°R	°F	°C
Zero thermal energy	0	0	−459.6	−273.15
Oxygen: liquid-gas	90.18	162.3	−297.3	−182.97
Water: solid-liquid	273.15	491.6	32	0
Water: liquid-gas	373.15	671.6	212	100
Gold: solid-liquid	1336.15	2405	1945.5	1063

temperature protection. A fail-safe mechanism may also be incorporated for system shutdown, when processing volatile or corrosive materials.

Summary

This chapter introduced the concepts of heat and temperature and their relationship to each other. The various temperature scales in use and the conversion equations between the scales are defined. The equations for heat transfer and heat storage are given. Temperature measuring instruments are described and their characteristics compared.

The highlights of this chapter on temperature and heat are as follows:

1. Temperature scales and their relation to each other are defined with examples on how to convert from one scale to the other
2. The transition of material between solid, liquid, and gaseous states or phase changes in materials when heat is supplied
3. The mechanism and equations of heat energy transfer and the effects of heat on the physical properties of materials
4. Definitions of the terms and standards used in temperature and heat measurements, covering both heat flow and capacity
5. The various temperature measuring devices including thermometers, bimetallic elements, pressure-spring devices, RTDs, and thermocouples
6. Considerations when selecting a temperature sensor for an application, thermal time constants, installation, and calibration

Problems

8.1 Convert the following temperatures to Fahrenheit: 115°C, 456 K, and 423°R.

8.2 Convert the following temperatures to Rankine: –13°C, 645 K, and –123°F.

8.3 Convert the following temperatures to Centigrade: 115°F, 356 K, and 533°R.

8.4 Convert the following temperatures to Kelvin: –215°C, –56°F, and 436°R.

8.5 How many calories of energy are required to raise the temperature of 3 ft^3 of water 15°F?

8.6 A 15-lb block of brass with a specific heat of 0.089 is heated to 189°F and then immersed in 5 gal of water at 66°F. What is the final temperature of the brass and water? Assume there is no heat loss.

8.7 A 4.3-lb copper block is heated by passing a direct current through it. If the voltage across the copper is 50 V and the current is 13.5 A, what will be the increase in the temperature of the copper after 17 min? Assume there is no heat loss.

8.8 A 129 kg lead block is heated to 176°C from 19°C, how many calories are required?

8.9 One end of a 9-in long × 7-in diameter copper bar is heated to 59.4°F, the far end of the bar is held at 23°C. If the sides of the bar are covered with thermal insulation, what is the rate of heat transfer?

8.10 On a winter's day the outside temperature of a 17-in thick concrete wall is −29°F, the wall is 15 ft long and 9 ft high. How many BTUs are required to keep the inside of the wall at 69°? Assume the thermal conductivity of the wall is 0.8 BTU/h ft°F.

8.11 When the far end of the copper bar in Prob. 8.9 has a 30-ft^2 cooling fin attached to the end of the bar and is cooled by air convection, the temperature of the fin rises to t °F. If the temperature of the air is 23°F and the coefficient of heat transfer of the surface is 0.22 BTU/h, what is the value of t?

8.12 How much heat is lost due to convection in a 25-min period from a 52 ft × 14-ft wall, if the difference between the wall temperature and the air temperature is 54°F and the surface of the wall has a heat transfer ratio of 0.17 BTU/h ft^2°F?

8.13 How much heat is radiated from a surface 1.5 ft × 1.9 ft if the surface temperatures is 125°F? The air temperature is 74°F and the radiation constant for the surface is 0.19×10^{-8} BTU/h ft^2°F^4?

8.14 What is the change in length of a 5-m tin rod if the temperature changes from 11 to 245°C?

8.15 The length of a 115-ft metal column changes its length to 115 ft 2.5 in when the temperature goes from −40 to 116°F. What is the coefficient of expansion of the metal?

8.16 A glass block measures 1.3 ft × 2.7 ft × 5.4 ft at 71°F. How much will the volume increase if the block is heated to 563°F?

8.17 What is the coefficient of resistance per degree Celsius of a material, if the resistance is 2246 Ω at 63°F and 3074 Ω at 405°F?

8.18 A tungsten filament has a resistance of 1998 Ω at 20°C. What will its resistance be at 263°C?

8.19 A chromel–alumel thermocouple is placed in a 1773°F furnace. Its reference is 67°F. What is the output voltage from the thermocouple?

8.20 A pressure-spring thermometer having a time constant of 1.7 s is placed in boiling water (212°F) after being at 69°F. What will be the thermometer reading after 3.4 s?

Chapter 9

Humidity, Density, Viscosity, and pH

Chapter Objectives

This chapter will introduce you to humidity, density, viscosity, and pH, and help you understand the units used in their measurement. This chapter will also familiarize you with standard definitions in use, and the instruments used for their measurement.

The salient points covered in this chapter are as follows:

- Humidity ratio, relative humidity, dew point, and its measurement
- Understanding and use of a psychrometric chart
- Instruments for measuring humidity
- Understand the difference between density, specific weight, and specific gravity
- Instruments for measuring density and specific gravity
- Definition of viscosity and measuring instruments
- Defining and measuring pH values

9.1 Introduction

Many industrial processes such as textiles, wood, chemical processing and the like, are very sensitive to humidity; consequently it is necessary to control the amount of water vapor present in these processes. This chapter discusses four physical parameters.

They are as follows:

1. Humidity
2. Density, specific weight, and specific gravity

3. Viscosity

4. pH values

9.2 Humidity

9.2.1 Humidity definitions

Humidity is a measure of the relative amount of water vapor present in the air or a gas.

Relative humidity (Φ) is the percentage of water vapor by weight present in a given volume of air or gas compared to the weight of water vapor present in the same volume of air or gas saturated with water vapor, at the same temperature and pressure, i.e.,

$$\text{Relative humidity} = \frac{\text{amount of water vapor present in a given volume of air or gas}}{\text{maximum amount of water vapor soluble in the same volume of air or gas } (p \text{ and } T \text{ constant})} \times 100 \quad (9.1)$$

An alternative definition using vapor pressures is as follows:

$$\text{Relative humidity} = \frac{\text{water vapor pressure in air or gas}}{\text{water vapor pressure in saturated air or gas } (T \text{ constant})} \times 100 \quad (9.2)$$

The term saturated means the maximum amount of water vapor that can be dissolved or absorbed by a gas or air at a given pressure and temperature. If there is any reduction of the temperature in saturated air or gas, water will condense out in the form of droplets, i.e., similar to mirrors steaming up when taking a shower.

Specific humidity, humidity ratio, or absolute humidity can be defined as the mass of water vapor in a mixture in grains (where 7000 grains = 1 lb) divided by the mass of dry air or gas in the mixture in pounds. The measurement units could also be in grams.

$$\text{Humidity ratio} = \frac{\text{mass of water vapor in a mixture}}{\text{mass of dry air or gas in the mixture}} \quad (9.3)$$

$$= \frac{\text{mass(water vapor)}}{\text{mass(air or gas)}} = \frac{0.622\, P(\text{water vapor})}{P(\text{mixture}) - P(\text{water vapor})} \quad (9.4)$$

$$= 0.622 \frac{P(\text{water vapor})}{P(\text{air or gas})} \quad (9.5)$$

where P(water vapor) is pressure and P(air or gas) is a partial pressure. The conversion factor between mass and pressure is 0.622.

Example 9.1 Examples of water vapor in the atmosphere are as follows:

Dark storm clouds (cumulonimbus) can contain 10 g/m^3 of water vapor.

Medium density clouds (cumulus congestus) can contain 0.8 g/m^3 of water vapor.

Light rain clouds (cumulus) contain 0.2 g/m^3 of water vapor.

Wispy clouds (cirrus) contain 0.1 g/m^3 of water vapor.

In the case of the dark storm clouds this equates to 100,000 tons of water vapor per square mile for a 10,000 ft tall cloud.

Dew point is the temperature at which condensation of the water vapor in air or a gas will take place as it is cooled at constant pressure, i.e., it is the temperature at which the mixture becomes saturated and the mixture can no longer dissolve or hold all of the water vapor it contains. The water vapor will now start to condense out of the mixture to form dew or a layer of water on the surface of objects present.

Dry-bulb temperature is the temperature of a room or mixture of water vapor and air (gas) as measured by a thermometer whose sensing element is dry.

Wet-bulb temperature is the temperature of the air (gas) as sensed by a moist element. Air is circulated around the element causing vaporization to take place; the heat required for vaporization (latent heat of vaporization) cools the moisture around the element, reducing its temperature.

Psychrometric chart is a somewhat complex combination of several simple graphs showing the relation between dry-bulb temperatures, wet-bulb temperatures, relative humidity, water vapor pressure, weight of water vapor per pound of dry air, BTUs per pound of dry air, and so on. While it may be a good engineering reference tool, it tends to overwhelm the student. For example Fig. 9.1 shows a psychrometric chart from Heat Pipe Technology, Inc. for standard atmospheric pressure; for other atmospheric pressures the sets of lines will be displaced.

To understand the various relationships in the chart it is necessary to break the chart down into only the lines required for a specific relationship.

Example 9.2 To obtain the relative humidity from the wet and dry bulb temperatures, the three lines shown in Fig. 9.2a should be used. These lines show the wet and dry bulb temperatures and the relative humidity lines. For instance, if the dry and wet bulb temperatures are measured as 76°F and 57°F, respectively, which when applied to Fig. 9.1 shows the two temperature lines as intersecting on the 30 percent relative humidity line, hence, the relative humidity is 30 percent. When the wet and dry bulb temperatures do not fall on a relative humidity line a judgment call has to be made for the value of the relative humidity.

Example 9.3 If the temperature in a room is 75°F and the relative humidity is 55 percent, how far can the room temperature drop before condensation takes place? Assume no other changes. In this case it is necessary to get the intersection of the dry bulb temperature and relative humidity lines, as shown in Fig. 9.2b and then the

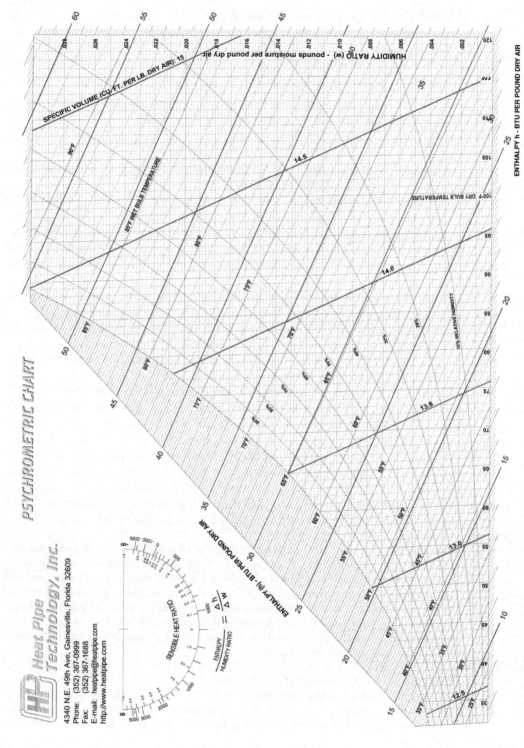

Figure 9.1 Psychrometric chart for air-water vapor mixtures. (*Courtesy of Heat Pipe Technology, Inc*)

Humidity, Density, Viscosity, and pH

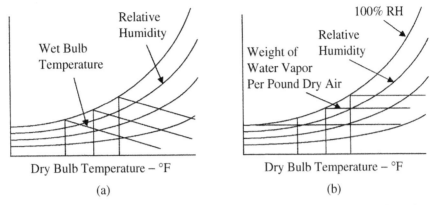

Figure 9.2 Lines required for finding (a) the relative humidity and (b) the condensation temperature.

corresponding horizontal line (weight of water vapor per pound of dry air). Using Fig. 9.1 the intersection falls on 0.01 lb of moisture per pound of dry air, because the weight of water vapor per pound of dry air will not change as the temperature changes. This horizontal line can be followed across to the left until it reaches the 100 percent relative humidity line (dew point). The temperature where these lines cross is the temperature where dew will start to form, i.e., 57°F. The wet and dry bulb temperatures are the same at this point. Note that in some charts the weight of water vapor in dry air is measured in grains, where 1 lb = 7000 grains or 1 grain = 0.002285 oz.

Example 9.4 This example compares the weight of water vapor in air at different humidity levels. The question is, how much more moisture does air at 80°F and 50 percent relative humidity contain than air at 60°F and 30 percent relative humidity? Using Fig. 9.3a as a reference, it is necessary to get the intersection of the dry bulb

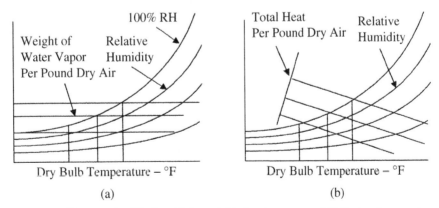

Figure 9.3 Lines required finding (a) the weight of water vapor and (b) the heating requirements.

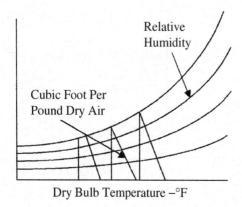

Figure 9.4 Lines required finding the volume per pound of dry air.

temperature and the relative humidity lines. The horizontal line where they intersect gives the weight of water vapor per pound of dry air as in the previous example. Using Fig. 9.1 the intersections are 0.0108 and 0.0032 lb, hence, the difference = 0.0076 lb or 53 grains.

Example 9.5 How much heat is required to raise the temperature of air at 50°F and 75 percent relative humidity to 75°F and 45 percent relative humidity? Referring to Fig. 9.3*b*, the intersection of the dry bulb temperature and relative humidity lines must be found, and hence, the total dry heat line passing through the intersection (these lines are an extension of the wet bulb temperature lines to the total heat per pound dry air scale). From Fig. 9.1 the intersection of the temperature and relative humidity lines fall on 18.2 and 27.2 Btus/lb of dry air, giving a difference of 9.0 Btus/lb of dry air.

Example 9.6 In air at 75°F and 45 percent relative humidity, how much space is occupied by a pound of dry air? The lines shown in Fig. 9.4 are used. The intersection of the dry bulb temperature and relative humidity lines on the cubic feet per pound of dry air line gives the space occupied by 1 lb of dry air. From Fig. 9.1 the lines intersect at 13.65 ft^3 giving this as the volume containing 1 lb of dry air.

9.2.2 Humidity measuring devices

Hygrometers. Devices that indirectly measure humidity by sensing changes in physical or electrical properties in materials due to their moisture content are called hygrometers. Materials such as hair, skin, membranes, and thin strips of wood change their length as they absorb water. The change in length is directly related to the humidity. Such devices are used to measure relative humidity from 20 to 90 percent, with accuracies of about ±5 percent. Their operating temperature range is limited to less than 70°C.

Laminate hygrometer is made by attaching thin strips of wood to thin metal strips forming a laminate. The laminate is formed into a helix as shown

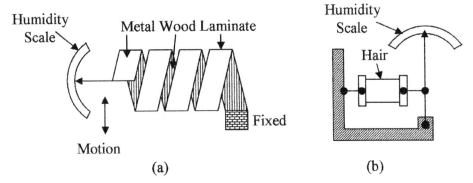

Figure 9.5 Two types of hygrometers using (a) metal/wood laminate and (b) hair.

in Fig. 9.5a, as the humidity changes the helix flexes due to the change in the length of the wood. One end of the helix is anchored, the other is attached to a pointer (similar to a bimetallic strip used in temperature measurements); the scale is graduated in percent humidity.

Hair hygrometer is the simplest and oldest type of hygrometer. It is made using hair as shown in Fig. 9.5b. Human hair lengthens by 3 percent when the humidity changes from 0 to 100 percent, the change in length can be used to control a pointer for visual readings or a transducer such as a linear variable differential transformers (LVDT) for an electrical output. The hair hygrometer has an accuracy of about 5 percent for the humidity range 20 to 90 percent over the temperature range 5 to 40°C.

Resistive hygrometer or resistive humidity sensors consist of two electrodes with interdigitated fingers on an insulating substrate as shown in Fig. 9.6a. The electrodes are coated with a hydroscopic material (one that absorbs water such as lithium chloride). The hydroscopic material provides a conductive path between the electrodes; the coefficient of resistance of the path is inversely proportional to humidity. Alternatively, the electrodes can be coated with a bulk polymer film that releases ions in proportion to the relative humidity; temperature

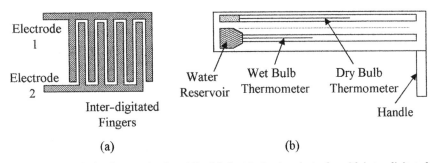

Figure 9.6 Methods of measuring humidity (a) electrical using electrodes with inter-digitated fingers covered with a hydroscopic material and (b) sling psychrometer.

correction can again be applied for an accuracy of 2 percent over the operating temperature range 40 to 70°C and relative humidity from 2 to 98 percent. An ac voltage is normally used with this type of device, i.e., at 1 kHz a relative humidity change from 2 to 98 percent will typically give a resistance change from 10 MΩ to 1 kΩ. Variations of this device are the electrolytic and the resistance-capacitance hygrometer.

Capacitive hygrometer. The dielectric constant of certain thin polymer films changes linearly with humidity, so that the capacitance between two plates using the polymer as the dielectric is directly proportional to humidity. The capacitive device has good longevity, a working temperature range of 0 to 100°C, a fast response time, and can be temperature compensated to give an accuracy of ±0.5 percent over the full humidity range.

Piezoelectric or sorption hygrometers use two piezoelectric crystal oscillators; one is used as a reference and is enclosed in a dry atmosphere, and the other is exposed to the humidity to be measured. Moisture increases the mass of the crystal which decreases its resonant frequency. By comparing the frequencies of the two oscillators, the humidity can be calculated. Moisture content of gases from 1 to 25,000 ppm can be measured.

Psychrometers. A psychrometer uses the latent heat of vaporization to determine the relative humidity. If the temperature of air is measured with a dry bulb thermometer and a wet bulb thermometer, the two temperatures can be used with a psychrometric chart to obtain the relative humidity, water vapor pressure, heat content, and weight of water vapor in the air. Water evaporates from the wet bulb trying to saturate the surrounding air. The energy needed for the water to evaporate cools the thermometer, so that the dryer the day, the more water evaporates and, hence, the lower the temperature of the wet bulb.

To prevent the air surrounding the wet bulb from saturating, there should be some air movement around the wet bulb. This can be achieved with a small fan or by using a sling psychrometer, which is a frame holding both the dry and wet thermometers that can rotate about a handle as shown in Fig. 9.6*b*. The thermometers are rotated for 15 to 20 s. The wet bulb temperature is taken as soon as rotation stops before it can change, and then the dry bulb temperature is taken (which does not change).

Dew point measuring devices. A simple method of measuring the humidity is to obtain the dew point. This is achieved by cooling the air or gas until water condenses on an object and then measuring the temperature at which condensation takes place. Typically, a mirrored surface, polished stainless steel, or silvered surface is cooled from the back side, by cold water, refrigeration, or Peltier cooling. As the temperature drops, a point is reached where dew from the air or gas starts to form on the mirror surface. The condensation is detected by the reflection of a beam of light by the mirror to a photocell. The intensity of the reflected light reduces as condensation takes place and the temperature of the mirror at that point can be measured.

Moisture content measuring devices. Moisture content of materials is very important in some processes. There are two methods commonly used to measure the moisture content; these are with the use of microwaves or by measuring the reflectance of the material to infrared rays.

Microwave absorption by water vapor is a method used to measure the humidity in a material. Microwaves (1 to 100 GHz) are absorbed by the water vapor in the material. The relative amplitudes of the transmitted and microwaves passing through a material are measured. The ratio of these amplitudes is a measure of the humidity content of the material.

Infrared absorption uses infrared rays instead of microwaves. The two methods are similar. In the case of infrared, the measurements are based on the ability of materials to absorb and scatter infrared radiation (reflectance). Reflectance depends on chemical composition and moisture content. An infrared beam is directed onto the material and the energy of the reflected rays is measured. The measured wavelength and amplitude of the reflected rays are compared to the incident wavelength and amplitude; the difference between the two is related to the moisture content.

Other methods of measuring moisture content are by color changes or by absorption of moisture by certain chemicals and measuring the change in mass, neutron reflection, or nuclear magnetic resonance.

Humidity application considerations. Although, wet and dry bulbs were the standard for making relative humidity measurements, more up to date and easier to make electrical methods such as capacitance and resistive devices are now available and will be used in practice. These devices are small, rugged, reliable, and accurate with high longevity, and if necessary can be calibrated by the National Institute of Standards and Technology (NIST) against accepted gravimetric hygrometer methods. Using these methods, the water vapor in a gas is absorbed by chemicals that are weighted before and after to determine the amount of water vapor absorbed in a given volume of gas from which the relative humidity can be calculated.

9.3 Density and Specific Gravity

9.3.1 Basic terms

The density, specific weight, and specific gravity were defined in Chap. 5 as follows:

Density ρ of a material is defined as the mass per unit volume. Units of density are pounds (slug) per cubic foot [lb (slug)/ft^3] or kilogram per cubic meter (kg/m^3).

Specific weight γ is defined as the weight per unit volume of a material, i.e., pounds per cubic foot (lb/ft^3) or newton per cubic meter (N/m^3).

Specific gravity (SG) of a liquid or solid is defined as the density of the material divided by the density of water or the specific weight of the material divided

TABLE 9.1 Density and Specific Weights

Material	Specific weight		Density		Specific gravity
	lb/ft^3	kN/m^3	slug/ft^3	×10^3 kg/m^3	
Acetone	49.4	7.74	1.53	0.79	0.79
Ammonia	40.9	6.42	1.27	0.655	0.655
Benzene	56.1	8.82	1.75	0.9	0.9
Gasoline	46.82	7.35	3.4	0.75	0.75
Glycerin	78.6	12.4	2.44	1.26	1.26
Mercury	847	133	26.29	13.55	13.55
Water	62.43	9.8	1.94	1.0	1.0

by the specific weight of water at a specified temperature. The specific gravity of a gas is its density/specific weight divided by the density/specific weight of air at 60°F and 1 atmosphere pressure (14.7 psia).

The relation between density and specific weight is given by

$$\gamma = \rho g \tag{9.6}$$

where g is the acceleration of gravity 32.2 ft/s^2 or 9.8 m/s^2 depending on the units being used.

Example 9.7 What is the density of a material whose specific weight is 27 kN/m^3?

$$\rho = \gamma/g = 27 \text{ kN/m}^3/9.8 \text{ m/s}^2 = 2.75 \times 10^3 \text{ kg/m}^3$$

Table 9.1 gives a list of the density and specific weight of some common materials.

9.3.2 Density measuring devices

Hydrometers are the simplest device for measuring the specific weight or density of a liquid. The device consists of a graduated glass tube, with a weight at one end, which causes the device to float in an upright position. The device sinks in a liquid until an equilibrium point between its weight and buoyancy is reached. The specific weight or density can then be read directly from the graduations on the tube. Such a device is shown in Fig. 9.7a.

Thermohydrometer is a combination of hydrometer and thermometer, so that both the specific weight/density and temperature can be recorded and the specific weight/density corrected from lookup tables for temperature variations to improve the accuracy of the readings.

Induction hydrometers are used to convert the specific weight or density of a liquid into an electrical signal. In this case, a fixed volume of liquid set by the overflow tube is used in the type of setup shown in Fig. 9.7b, the displacement device, or hydrometer, has a soft iron or similar metal core attached. The core is positioned in a coil which forms part of a bridge circuit. As the density/specific

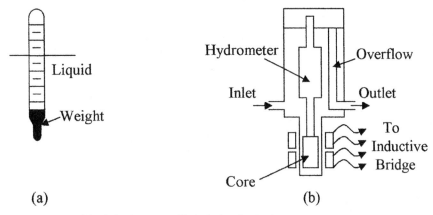

Figure 9.7 (a) A basic hydrometer, (b) An induction hydrometer.

weight of the liquid changes, the buoyant force on the displacement device changes. This movement can be measured by the coil and converted into a density reading.

Vibration sensors are an alternate method of measuring the density of a fluid (see Fig. 9.8a). Fluid is passed through a U tube which has a flexible mount so that it can vibrate when driven from an outside source. The amplitude of the vibration decreases as the specific weight or density of the fluid increases, so that by measuring the vibration amplitude the specific weight/density can be calculated.

Pressure at the base of a column of liquid of known height (h) can be measured to determine the density and specific gravity of a liquid. The density of the liquid is given by

$$\rho = \frac{p}{gh} \tag{9.7}$$

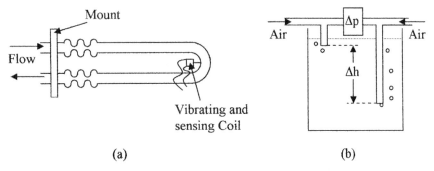

Figure 9.8 Alternative methods for density measurement are (a) vibration sensor and (b) bubbler system.

The specific weight is given by

$$\rho = \frac{p}{h} \tag{9.8}$$

Example 9.8 What is the pressure at the base of a column of liquid if the height of the column is 298 cm and the density of the liquid is 1.26×10^3 kg/m^3?

$$p = \rho g h = 1.26 \times 9.8 \times 298/100 = 36.8 \text{ Pa}$$

The weight of a known volume of the liquid can be used to determine density, i.e., a container of known volume can be filled with a liquid and weighted full and empty. The difference in weight gives the weight of liquid, from which the density can be calculated using the following equation:

$$\rho = \frac{W_f - W_c}{g \times \text{Vol}} \tag{9.9}$$

where W_f = weight of container + liquid
W_c = weight of container
Vol = volume of the container

Differential bubblers can be used to measure liquid density or specific weight. Figure 9.8*b* shows the setup using a bubbler system. Two air supplies are used to supply two tubes whose ends are at different depths in a liquid, the difference in air pressures between the two air supplies is directly related to the density of the liquid by the following equation:

$$\rho = \frac{\Delta p}{g \Delta h} \tag{9.10}$$

where Δp is the difference in the pressures and Δh the difference in the height of the bottoms of the two tubes.

Example 9.9 What is the density of a liquid in a bubbler system if pressures of 500 Pa and 23 kPa are measured at depths of 15 cm and 6.5 m, respectively?

$$\rho = \frac{23 - \dfrac{500}{1000}}{\left(6.5 - \dfrac{15}{100}\right) \times 9.8} = \frac{23 - 0.5}{(6.5 - 0.15) \times 9.8} = 0.36 \times 10^3 \text{ kg/m}^3$$

Radiation density sensors consist of a radiation source located on one side of a pipe or container and a sensing device on the other. The sensor is calibrated with the pipe or container empty, and then filled. Any difference in the measured radiation is caused by the density of the liquid which can then be calculated.

Gas densities are normally measured by sensing the frequency of vibration of a vane in the gas, or by weighing a volume of the gas and comparing it to the weight of the same volume of air.

9.3.3 Density application considerations

Ideally, when measuring the density of a liquid, there should be some agitation to ensure uniform density throughout the liquid. This is to avoid density gradients due to temperature gradients in the liquid and incomplete mixing of liquids at different temperatures. Excessive agitation should be avoided.

Density measuring equipment is available for extreme temperatures and pressures, i.e., from 150 to 600°F and for pressures in excess of 1000 psi. When measuring corrosive, abrasive, volatile liquids, and the like, radiation devices should be considered.

9.4 Viscosity

Viscosity was introduced in Chap. 7; it will be discussed in this chapter in more detail.

9.4.1 Basic terms

Viscosity μ in a fluid is the resistance to its change of shape, which is due to molecular attraction in the liquid that resists any change due to flow or motion. When a force is applied to a fluid at rest, the molecular layers in the fluid tend to slide on top of each other as shown in Fig. 9.9a. The force F resisting motion in a fluid is given by

$$F = \frac{\mu A V}{y} \qquad (9.11)$$

where A = boundary area being moved
V = velocity of the moving boundaries
y = distance between boundaries
μ = coefficient of viscosity, or dynamic viscosity

The units of measurement must be consistent.

Sheer stress τ is the force per unit area and is given in the following formula:

$$\mu = \frac{\tau y}{V} \qquad (9.12)$$

where τ is the shear stress or force per unit area.

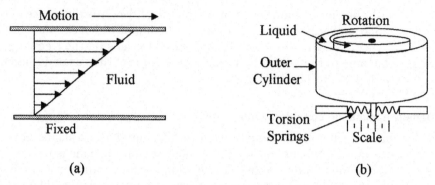

Figure 9.9 Illustration of (a) Newtonian laminar flow and (b) a drag-type viscometer.

If F is in pounds, A in square feet, V in feet per second, and y in feet, then μ is in pound seconds per square feet. Whereas, if F is in newton, A in square meter, V in meters per second, and y in meters, then μ is in newton seconds per square meter. A sample list of fluid viscosities is given in Table 9.2.

The standard unit of viscosity is the poise, where a centipoise (poise/100) is the viscosity of water at 68.4°F. Conversions are given in Table 7.1. (1 centipoise = 2.09×10^{-5} lb s/ft^2).

When the temperature of a body increases, more energy is imparted to the atoms making them more active and thus effectively reducing the molecular attraction. This in turn reduces the attraction between the fluid layers lowering the viscosity, i.e., viscosity decreases as temperature increases.

Newtonian fluids are fluids that exhibit only laminar flow as shown in Fig. 9.9a and are consistent with temperature. Only newtonian fluids will be considered. Non–Newtonian fluid dynamics is complex and considered to be outside the scope of this text.

9.4.2 Viscosity measuring instruments

Viscometers or viscosimeters are used to measure the resistance to motion of liquids and gases. Several different types of instruments have been designed to measure viscosity, such as the inline falling-cylinder viscometer, the drag-type

TABLE 9.2 Dynamic Viscosities, at 68°F and Standard Atmospheric Pressure

Fluid	μ(lb·s/ft^2)	Fluid	μ(lb·s/ft^2)
Air	38×10^{-8}	Carbon dioxide	31×10^{-8}
Hydrogen	19×10^{-8}	Nitrogen	37×10^{-8}
Oxygen	42×10^{-8}	Carbon tetrachloride	20×10^{-6}
Ethyl alcohol	25×10^{-6}	Glycerin	18×10^{-3}
Mercury	32×10^{-6}	Water	21×10^{-6}
Water	1×10^{-2} poise		

viscometer, and the Saybolt universal viscometer. The rate of rise of bubbles in a liquid can also be used to give a measure of the viscosity of a liquid.

The falling-cylinder viscometer uses the principle that an object when dropped into a liquid will descend to the bottom of the vessel at a fixed rate; the rate of descent is determined by the size, shape, density of the object, and the density and viscosity of the liquid. The higher the viscosity, the longer the object will take to reach the bottom of the vessel. The falling-cylinder device measures the rate of descent of a cylinder in a liquid and correlates the rate of descent to the viscosity of the liquid.

Rotating disc viscometer is a drag-type device. The device consists of two concentric cylinders and the space between the two cylinders is filled with the liquid being measured, as shown in Fig. 9.9b. The inner cylinder is driven by an electric motor and the force on the outer cylinder is measured by noting its movement against a torsion spring; the viscosity of the liquid can then be determined.

The *Saybolt instrument* measures the time for a given amount of fluid to flow through a standard size orifice or a capillary tube with an accurate bore. The time is measured in Saybolt seconds, which is directly related and can be easily converted to other viscosity units.

Example 9.10 Two parallel plates separated by 0.45 in are filled with a liquid with a viscosity of 7.6×10^{-4} lb·s/ft^2. What is the force acting on 1 ft^2 of the plate, if the other plate is given a velocity of 4.4 ft/s?

$$F = \frac{7.6 \times 10^4 \text{ lb·s} \times 1 \text{ ft}^2 \times 4.4 \text{ ft} \times 12}{\text{ft}^2 \times 0.45 \text{ ft·s}} = 0.089 \text{ lb}$$

9.5 pH Measurements

9.5.1 Basic terms

In many process operations, pure and neutral water is required for cleaning or diluting other chemicals, i.e., the water is not acidic or alkaline. Water contains both hydrogen ions and hydroxyl ions. When these ions are in the correct ratio the water is neutral. An excess of hydrogen ions causes the water to be acidic and when there is an excess of hydroxyl ions, the water is alkaline. The pH (power of hydrogen) of the water is a measure of its acidity or alkalinity; neutral water has a pH value of 7 at 77°F (25°C). When water becomes acidic the pH value decreases. Conversely, when the water becomes alkaline the pH value increases. The pH values use a log to the base 10 scale, i.e., a change of 1 pH unit means that the concentration of hydrogen ions has increased (or decreased) by a factor of 10 and a change of 2 pH units means the concentration has changed by a factor of 100. The pH value is given by

$$\text{pH} = \log_{10}[1/\text{hydrogen ion concentration}] \quad (9.13)$$

The pH value of a liquid can range from 0 to 14. The hydrogen ion concentration is in grams per liter, i.e., a pH of 4 means that the hydrogen ion concentration is 0.0001 g/l at 25°C. Strong hydrochloric or sulfuric acids will have a pH of 0 to 1.

4 % caustic soda	pH = 14
Lemon and orange juice	pH = 2 to 3
Ammonia	pH is about 11

Example 9.11 The hydrogen ion content in water goes from 0.15 g/L to 0.0025 g/L. How much does the pH change?

$$pH_1 = \log\left(\frac{1}{0.15}\right) = 0.824$$

$$pH_2 = \log\left(\frac{1}{0.0025}\right) = 2.6$$

Change in pH = 0.824 − 2.6 = −1.776

9.5.2 pH measuring devices

The pH is normally measured by chemical indicators or by pH meters. The final color of chemical indicators depends on the hydrogen ion concentration; their accuracy is only 0.1 to 0.2 pH units. For indication of acid, alkali, or neutral water, litmus paper is used; it turns pink when acidic, blue when alkaline, and stays white if neutral.

A pH sensor normally consists of a sensing electrode and a reference electrode immersed in the test solution which forms an electrolytic cell, as shown in Fig. 9.10a. One electrode contains a saturated potassium chloride (alkaline) solution to act as a reference; the electrode is electrically connected to the test solution via the liquid junction. The other electrode contains a buffer which sets the electrode in contact with the liquid sample. The electrodes are connected to a differential amplifier, which amplifies the voltage difference between the electrodes, giving an output voltage that is proportional to the pH of the solution. Figure 9.10b shows the pH sensing electrode.

9.5.3 pH application considerations

The pH of neutral water varies with temperature, i.e., neutral water has a pH of about 7.5 at 32°F and about 6 at 212°F. pH systems are normally automatically temperature compensated. pH test equipment must be kept clean and free from contamination. Calibration of test equipment is done with commercially available buffer solutions with known pH values. Again, cleaning between each reading is essential to prevent contamination.

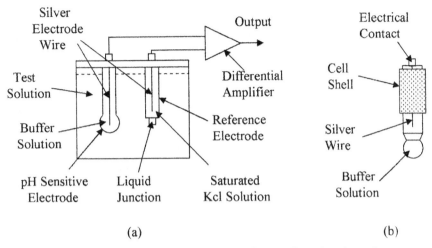

Figure 9.10 Shows the set up of (a) a pH sensor and (b) a pH sensing electrode.

Summary

This chapter introduced humidity, its relation to dew point, and temperature. The psychrometric chart is shown and instructions given on how to read the chart. Humidity terms and instruments are described. Density, specific weight, and specific gravity are defined together with measurement instruments. Viscosity and pH were introduced with definitions and measuring instruments.

The prime points discussed in this chapter were as follows:

1. The definition of and relationship between specific humidity, relative humidity, and dew point
2. Use of the psychrometric chart for obtaining dew point and the weight of water vapor dissolved in the atmosphere from temperature data
3. The various types of instruments used for the direct and indirect measurement of humidity
4. Density, specific weight, and specific gravity are defined with examples in both English and SI units. Instruments for measuring specific weight and specific gravity are given
5. Instruments used in the measure of density and specific weight
6. The basic terms and definitions used in viscosity, its relation to flow with examples, and instruments used for measuring viscosity
7. An introduction to pH terms, its value when determining acidity or alkalinity, and the instruments used to measure pH.

Chapter Nine

Problems

9.1 The dry bulb of a wet/dry thermometer reads 120°F. What is the relative humidity if the wet bulb reads (a) 90°F, (b) 82°F, (c) 75°F?

9.2 The wet bulb of a wet/dry thermometer reads 85°F. What will the dry bulb read if the relative humidity is (a) 80 percent, (b) 60 percent, (c) 30 percent?

9.3 What is the relative humidity if the wet and dry bulbs in a wet/dry thermometer read 75°F and 85°F respectively?

9.4 The dry bulb of a wet/dry thermometer reads 60°F, and the relative humidity is 47 percent. What will the wet bulb read and what is the absolute humidity?

9.5 The wet and dry bulbs of a wet/dry thermometer read 75°F and 112°F, respectively. What is the relative and absolute humidity?

9.6 If the air temperature is 85°F, what is the water vapor pressure corresponding to a relative humidity of 55 percent and at 55°F with 85 percent relative humidity?

9.7 How much water is required to raise the relative humidity of air from 25 to 95 percent if the temperature is held constant at 95°F?

9.8 How much heat and water are added per pound of dry air to increase the relative humidity from 15 to 80 percent with a corresponding temperature increase from 42 to 95°F?

9.9 How much water does dry air contain at 105°F and 55 percent relative humidity?

9.10 How much heat is required to heat one pound of dry air from 35 to 80°F if the relative humidity is constant at 80 percent?

9.11 How much space is occupied by 4.7 lb of dry air if the air temperature is 80°F and the relative humidity is 82 percent?

9.12 The two tubes in a bubbler system are placed in a liquid with a density of 1.395 slugs/ft^3. If the bottom ends of the bubbler tubes are 3.5 and 42.7 in below the surface of the liquid, what is the differential pressure supplied to the two bubblers?

9.13 A tank is filled to a depth of 54 ft with liquid having a density of 1.234 slugs/ft^3. What is the pressure on the bottom of the tank?

9.14 What would be the SG of a gas with a specific weight of 0.127 lb/ft^3, if the density of air under the same conditions is 0.0037 slugs/ft^3?

9.15 A square plate 1.2 ft on a side is centrally placed in a channel 0.23-in wide filled with a liquid with a viscosity of 7.3×10^{-5} lb s/ft^2. If the plate is 0.01-in thick what force is required to pull the plate along the channel at 14.7 ft/s?

9.16 Two parallel plates 35 ft^2, 1.7-in apart are placed in a liquid with a viscosity of 2.1×10^{-4} lb s/ft^2. If a force of 0.23 lb is applied to one plate in a direction parallel to the plates with the other plate fixed, what is the velocity of the plate?

9.17 What is the pH of a solution, if there is a concentration of 0.0006 g/L of hydrogen ions?

9.18 What is the change in hydrogen concentration factor if the pH of a solution changes from 3.5 to 0.56?

9.19 What is the concentration of hydrogen ions if the pH is 13.2?

9.20 What is the hydrogen concentration of a neutral solution?

Chapter 10
Other Sensors

Chapter Objectives

This chapter will help you understand and become familiar with other sensors that play a very important part in process control, but may not be encountered on a daily basis. The following are covered in this chapter:

- Position, distance, velocity, and acceleration sensors
- Rotation sensors using light and Hall effect sensors
- Force, torque, load cells, and balances
- Smoke detectors, gas, and chemical sensors in industry
- Sound and light measurements
- Sound and optical devices

10.1 Introduction

There are many sensors other than level, pressure, flow, and temperature that may not be encountered on a day to day basis—such as position, force, smoke, and chemical sensors—but play an equally important part in process control in today's high-technology industries and/or for operator protection. These sensors will not be discussed in as much detail as the sensors already discussed. However, the student should be aware of their existence and operation.

10.2 Position and Motion Sensing

10.2.1 Basic position definitions

Many industrial processes require both linear and angular position and motion measurements. These are required in robotics, rolling mills, machining operations, numerically controlled tool applications, and conveyers. In some applications it is also necessary to measure speed, acceleration, and vibration. Some transducers use position sensing devices to convert temperature and/or pressure into electrical units and controllers can use position sensing devices to monitor the position of an adjustable valve for feedback control.

Absolute position is the distance measured with respect to a fixed reference point and can be measured whenever power is applied.

Incremental position is a measure of the change in position and is not referenced to a fixed point. If power is interrupted, the incremental position change is lost. An additional position reference such as a limit switch is usually used with this type of sensor. This type of sensing can give very accurate positioning of one component with respect to another and is used when making master plates for tooling and the like.

Rectilinear motion is measured by the distance traversed in a given time, velocity when moving at a constant speed, or acceleration when the speed is changing in a straight line.

Angular position is a measurement of the change in position of a point about a fixed axis measured in degrees or radians, where one complete rotation is 360° or 2π radians. The degrees of rotation of a shaft can be absolute or incremental. These types of sensors are also used in rotating equipment to measure rotation speed as well as shaft position and to measure torque displacement.

Arc-minute is an angular displacement of 1/60 of a degree.

Angular motion is a measure of the rate of rotation. Angular velocity is a measure of the rate of rotation when rotating at a constant speed about a fixed point or angular acceleration when the rotational speed is changing.

Velocity or speed is the rate of change of position. This can be a linear measurement, i.e., feet per second (ft/s), meters per second (m/s), and so forth, or angular measurement, i.e., degrees per second, radians per second, rate per minute (r/m), and so forth.

Acceleration is the rate of change of speed, i.e., feet per second squared (ft/s^2), meters per second squared (m/s^2), and the like for linear motion, or degrees per second squared, radians per second squared, and the like, in the case of rotational motion.

Vibration is a measure of the periodic motion about a fixed reference point or the shaking that can occur in a process due to sudden pressure changes, shock, or unbalanced loading in rotational equipment. Peak accelerations of 100 g can occur during vibrations which can lead to fracture or self destruction. Vibration sensors are used to monitor the bearings in heavy rollers such as those used in rolling mills; excessive vibration indicates failure in the bearings or damage to rotating parts that can then be replaced before serious damage occurs.

10.2.2 Position and motion measuring devices

Potentiometers are a convenient method of converting the displacement in a sensor to an electrical variable. The wiper or slider arm of a linear potentiometer can be mechanically connected to the moving section of a sensor. Where rotation is involved, a single or multiturn (up to 10 turns) rotational type of potentiometer can be used. For stability, wire-wound devices should be used, but in environmentally-unfriendly conditions, lifetime of the potentiometer may be limited by dirt, contamination, and wear.

Linear variable differential transformers (LVDT) are devices that are used for measuring small distances and are an alternative to the potentiometer. The device consists of a primary coil with two secondary windings one on either side of the primary. (see Fig. 10.1a). A movable core when centrally placed in the primary will give equal coupling to each of the secondary coils. When an ac voltage is applied to the primary, equal voltages will be obtained from the secondary windings which are wired in series opposition to give zero output voltage, as shown in Fig. 10.1b. When the core is slightly displaced an output voltage proportional to the displacement will be obtained. These devices are not as cost effective as potentiometers but have the advantage of being noncontact. The outputs are electrically isolated, accurate, and have better longevity than potentiometers.

Light interference lasers are used for very accurate incremental position measurements. Monochromatic light (single frequency) can be generated with a laser and collimated into a narrow beam. The beam is reflected by a mirror attached to the moving object which generates interference fringes with the incident light as it moves. The fringes can be counted as the mirror moves. The wavelength of the light generated by a laser is about 5×10^{-7} m, so relative positioning to this accuracy over a distance of $\frac{1}{2}$ to 1 m is achievable.

Ultrasonic, infrared, laser, and microwave devices can be used for distance measurement. The time for a pulse of energy to travel to an object and be reflected back to a receiver is measured, from which the distance can be calculated, i.e.,

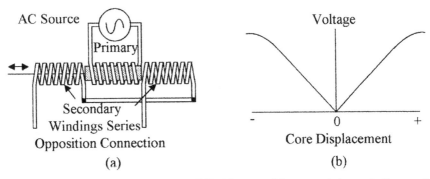

Figure 10.1 Demonstrated is (*a*) the LVDT with a movable core and three windings and (*b*) the secondary voltage versus core displacement for the connections shown.

the speed of ultrasonic waves is 340 m/s and the speed of light and microwaves is 3×10^8 m/s. Ultrasonic waves can be used to measure distances from 1 to about 50 m, whereas the light and microwaves are used to measure longer distances.

If an object is in motion the Doppler effect can be used to determine its speed. The Doppler effect is the change in frequency of the reflected waves caused by the motion of the object. The difference in frequency between the transmitted and reflected signal can be used to calculate the velocity of the object.

Hall effect sensors detect changes in magnetic field strength and are used as a close proximity protector. The Hall effect occurs in semiconductor devices and is shown in Fig. 10.2a. Without a magnetic field the current flows directly through the semiconductor plate and the Hall voltage is zero. Under the influence of a magnet field, as shown, the current path in the semiconductor plate becomes curved, giving a Hall voltage between the sides adjacent to the input/output current. In Fig. 10.2b a Hall effect device is used to detect the rotation of a cog wheel. As the cogs move pass the Hall device, the strength of the magnetic field is greatly enhanced causing an increase in the Hall voltage. The device can be used to measure linear as well as rotational position or speed and can also be used as a limit switch.

Magneto resistive element (MRE) is an alterative to the Hall effect device. In the case of the MRE its resistance changes with magnetic field strength.

Optical devices detect motion by sensing the presence or absence of light. Figure 10.3 shows two types of optical discs used in rotational sensing. Figure 10.3a shows an incremental optical shaft encoder. Light from the light-emitting diode (LED) shines through windows in the disc on to an array of photodiodes. As the shaft turns, the position of the image moves along the array of diodes. At the end of the array, the image of the next slot is at the start of the array. The relative position of the wheel with respect to its previous location can be obtained by counting the number of photodiodes traversed and multiplying them by the number of slots monitored. The diode array enhances the accuracy of the position of the slots, i.e., the resolution of the sensor is 360° divided by the number

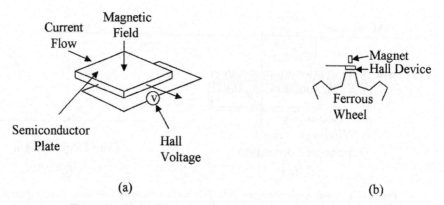

Figure 10.2 Shown is a semiconductor plate used as (*a*) Hall effect device and (*b*) application of a Hall effect device for measuring the speed and position of a cog wheel.

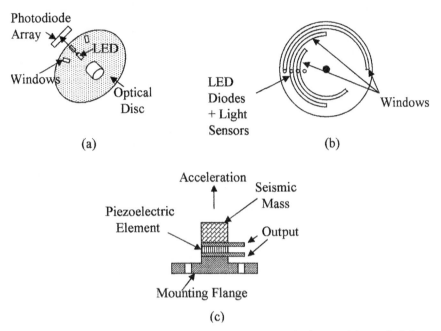

Figure 10.3 Shows (*a*) an incremental optical disc, (*b*) an absolute position optical discs, and (*c*) a piezoelectric accelerometer.

of slots in the disc divided by the number of diodes in the array. The slots can also be replaced by reflective strips, in which case the light from the LED is reflected back to a photodiode array.

Only one slot in the disc is required to measure rate per minute. Figure 10.3*b* shows an absolute position encoder. An array of LEDs (one for each window) with a corresponding photo detector for each window can give the position of the wheel at any time. Only three windows are shown in the figure, for greater accuracy more slots would be used. The pattern shown on the disc is for the gray code. Other patterns may be used on the disc such as the binary code.

Optical devices have many uses in industry other than for the measurement of the position and speed of rotating equipment. Optical devices are used for counting objects on conveyer belts on a production line, measurement and control of the speed of a conveyer belt, location and position of objects on a conveyer, location of registration marks for alignment, bar code reading, measurement and thickness control, and detecting for breaks in filaments and so forth.

Power lasers can also be included with optical devices as they are used for scribing and machining of metals, laminates, and the like.

Accelerometers sense speed changes by measuring the force produced by the change in velocity of a known mass (seismic mass), see Eq. (10.1). These devices can be made with a cantilevered mass and a strain gauge for force measurement or can use capacitive measurement techniques. Accelerometers are now commercially available, made using micromachining techniques. The devices can be as small as 500 µm × 500 µm, so that the effective loading by the accelerometer on a measurement is very small. The device is a small cantilevered seismic mass that uses capacitive changes to monitor the position of the mass. Piezoelectric devices similar to the one shown in Fig. 10.3c are also used to measure acceleration. The seismic mass produces a force on the piezoelectric element during acceleration which causes a voltage to be developed across the element. Accelerometers are used in industry for the measurement of changes in velocity of moving equipment, in the automotive industry as crash sensors for air bag deployment, and in shipping crates where battery operated recorders are used to measure shock during the shipment of expensive and fragile equipment.

Vibration sensors typically use acceleration devices to measure vibration. Micromachined accelerometers make good vibration sensors for frequencies up to about 1 kHz. Piezoelectric devices make good vibration sensors with an excellent high-frequency response for frequencies up to 100 kHz. These devices have very low mass so that the damping effect is minimal. Vibration sensors are used for the measurement of vibration in bearings of heavy equipment and pressure lines.

10.2.3 Position application consideration

Optical position sensors require clean operating conditions and in dirty or environmentally-unfriendly applications they are being replaced by Hall or MRE devices in both rotational and linear applications. These devices are small, sealed, and rugged with very high longevity and will operate correctly in fluids, in a dirty environment, or in contaminated areas.

Optical devices can be used for reading bar codes on containers and imaging. Sensors in remote locations can be powered by solar cells that fall into the light sensor category.

10.3 Force, Torque, and Load Cells

10.3.1 Basic definitions of force and torque

Many applications in industry require the measurement of force or load. Force is a vector and acts in a straight line, it can be through the center of a mass, or be offset from the center of the mass to produce a torque, or with two forces a couple. Force can be measured with devices such as strain gages. In other applications where a load or weight is required to be measured the sensor can be a load cell.

Mass is a measure of the quantity of material in a given volume of an object.

Force is a term that relates the mass of an object to its acceleration and acts through its center of mass, such as the force required to accelerate a mass at a given rate. Forces are defined by magnitude and direction and are given by the following:

$$\text{Force } (F) = \text{mass } (m) \times \text{acceleration } (a) \tag{10.1}$$

Example 10.1 What force is required to accelerate a mass of 27 kg at 18 m/s²?

$$\text{Force} = 27 \times 18 \text{ N} = 486 \text{ N}$$

Weight of an object is the force on a mass due to the pull of gravity, which gives the following:

$$\text{Weight } (w) = \text{mass } (m) \times \text{gravity } (g) \tag{10.2}$$

Example 10.2 What is the mass of a block of metal that weighs 29 lb?

$$\text{Mass} = 29/32.2 = 0.9 \text{ lb (slug)}$$

Torque occurs when a force acting on a body tends to cause the body to rotate and is defined by the magnitude of the force times the perpendicular distance from the line of action of the force to the center of rotation (see Fig. 10.4a). Units of torque are pounds (lb), feet (ft), or newton meter (N·m). Torque is sometimes referred to as the moment of the force, and is given by

$$\text{Torque } (t) = F \times d \tag{10.3}$$

A *couple* occurs when two parallel forces of equal amplitude, but in opposite directions, are acting on an object to cause rotation, as shown in Fig. 10.4b and is given by the following equation:

$$\text{Couple } (c) = F \times d \tag{10.4}$$

10.3.2 Force and torque measuring devices

Force and weight can be measured by comparison as in a lever-type balance which is an ON/OFF system. A spring balance or load cell can be used to generate an electrical signal that is required in most industrial applications.

Analytical or lever balance is a device that is simple and accurate, and operates on the principle of torque comparison. Figure 10.4c shows a diagram of a balance.

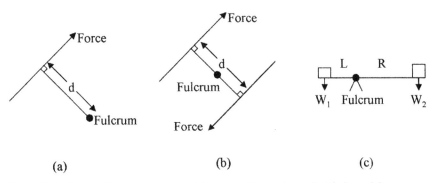

Figure 10.4 Types of forces shown are (a) torque, (b) couple, and (c) balanced forces.

When in balance the torque on one side of the fulcrum is equal to the torque on the other side of the fulcrum, from which we get the following:

$$W_1 \times L = W_2 \times R \qquad (10.5)$$

where W_1 is a weight at a distance L from the fulcrum and W_2 the counter balancing weight at a distance R from the fulcrum.

Example 10.3 Two pounds of potatoes are being weighted with a balance, the counter weight on the balance is 0.5 lb. If the balance arm from the potatoes to the fulcrum is 6 in long, how far from the fulcrum must the counter balance be placed?

$$2 \text{ lb} \times 0.5 \text{ ft} = 0.5 \text{ lb} \times d \text{ ft}$$
$$d = 2 \times 0.5/0.5 = 2 \text{ ft}$$

Spring transducer is a device that measures weight by measuring the deflection of a spring when a weight is applied, as shown in Fig. 10.5a. The deflection of the spring is proportional to the weight applied (provided the spring is not stressed), according to the following equation:

$$F = Kd \qquad (10.6)$$

where F = force in pounds or newtons
K = spring constant in pounds per inch or newtons per meter
d = spring deflection in inches or meters

Example 10.4 When a container is placed on a spring balance with an elongation constant of 65 lb/in (11.6 kg/cm) the spring stretches 3.2 in (8.1 cm). What is the weight of the container?

$$\text{Weight} = 65 \text{ lb/in} \times 3.2 \text{ in} = 208 \text{ lb}$$
$$= 11.6 \text{ kg/cm} \times 8.1 \text{ cm} = 93.96 \text{ kg}$$

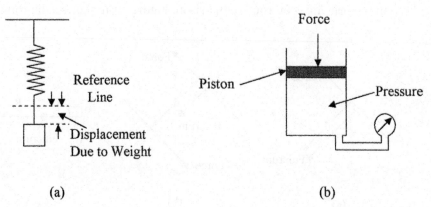

Figure 10.5 Force measuring devices (a) spring balance and (b) using pressure to measure force.

Hydraulic and pneumatic devices can be used to measure force. This can be done by monitoring the pressure in a cylinder when the force (pounds or newtons) is applied to a piston as shown in Fig. 10.5b, the relation between force (F) and pressure (p) is given by

$$F = pA \qquad (10.7)$$

where A is the area of the piston.

Example 10.5 What is the force acting on a 14-in (35.6-cm) diameter piston, if the pressure gage reads 22 psi (152 kPa)?

$$\text{Force} = \frac{22 \times 3.14 \times 14^2}{4} = 3385 \text{ lb}$$

$$= \frac{152 \times 3.14 \times 35.6 \times 35.6}{4 \times 100 \times 100} = 15.1 \text{ kN}$$

Piezoelectric devices, as previously noted, produce an electrical charge between the opposite faces of a crystal when the crystal is deformed by a force that makes them suitable for use as a force sensor (see Fig. 10.3c). Many crystals exhibit the piezoelectric effect. Some common crystals are as follows:

Quartz

Rochelle salt

Lithium sulphate

Tourmaline

Quartz devices have good sensitivity but have high output impedance. The output voltage drifts under low loading due to noise and temperature effects, but is well suited for measuring rapidly changing forces as well as static forces.

Tensile and compressive forces are measured with strain gauges; a strain gauge can use a piezoresistive material or other types of material that changes their resistance under strain. Figure 10.6a shows the use of a strain gauge to measure the strain in a solid body under stress from a tensile force, in this case the material under tension elongates and narrows. Strain gauges, as shown, are used to measure stress in a material from which the properties of the material can be calculated. A strain gauge can be used to measure stress from compressive forces as shown in Fig. 10.6b. An object under compressive forces will shorten and fatten and the strain can be measured.

Weight measurements are made with load cells which can be capacitive, electromagnetic, use piezoelectric elements, or strain gauges. A capacitive load cell is shown in Fig. 10.6c. The capacitance is measured between a fixed plate and a diaphragm. The diaphragm moves towards the fixed plate when force or pressure is applied, giving a capacitive change proportional to the force.

Dynamometer is a device that uses the twist or bending in a shaft due to torque to measure force. One such device is the torque wrench used to tighten

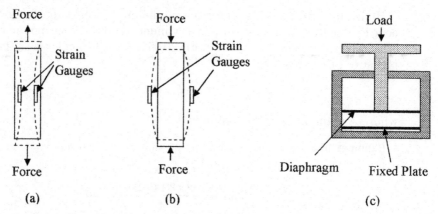

Figure 10.6 Examples of (*a*) a solid object under tensile strain, (*b*) a solid body under compression, and (*c*) a capacitive load cell.

bolts to a set level, which can be required in some valve housing. The allowable torque for correct assembly will be given in the manufacturer's specification. The twist in a shaft from a motor can be used to measure the torque output from the motor.

10.3.3 Force and torque application considerations

In most applications compensation should be made for temperature effects. Electrical transducers can be compensated by using them in a bridge circuit with a compensating device in the adjacent arm of the bridge. Changes in material characteristics due to temperature changes can be compensated using temperature sensors and applying a correction factor to the measurement. Vibration can also be a problem when measuring force, but this can usually be corrected by damping the movement of the measuring system.

10.4 Smoke and Chemical Sensors

The detection of smoke, radiation, and chemicals is of great importance in industrial processing not only as it relates to the safety of humans and the control of environment pollution both atmospheric and ground, but is also used in process control applications to detect the presence, absence, or levels of impurities in processing chemicals.

Smoke detectors and heat sensors (automatic sprinklers) are now commonplace in industry for the protection of people, equipment, and monitoring and control of chemical reactions. Low-cost smoke detectors using infrared sensing or ionization chambers are commercially available. Many industrial processes use a variety of gases in processing–such as inert gases (nitrogen)—to prevent contamination from oxygen in the air, or conversely, gases or chemicals can be introduced to give a desired reaction. It is therefore necessary to be able to monitor, measure, and control a wide variety of gases and chemicals. A wide variety of gas and chemical sensors are available. Of these, the Taguchi-type of sensor is one of the more common types of sensors.

10.4.1 Smoke and chemical measuring devices

Infrared sensors detect changes in the signal received from a light emitting diode due to the presence of smoke in the light path or the presence of an object in the light path.

Ionization chambers are devices that detect the leakage current between two plates that have a voltage between them. The leakage occurs when carbon particles from smoke are present and provide a conductive path between the plates.

Taguchi-type sensors are used for the detection of hydrocarbon gases, such as carbon monoxide and dioxide, methane, and propane. The Taguchi sensor has an element coated with an oxide of tin that combines with hydrocarbon to give a change in electrical resistance which can be detected. Periodically the element is heated and the chemical reaction is reversed, reducing the coating back to tin oxide. Likewise, the sensing process can be repeated. The tin oxide can be made sensitive to different hydrocarbons by using different oxides of tin, different deposition techniques, and so on.

10.4.2 Smoke and chemical application consideration

Many hazardous, corrosive, toxic, and environmentally-unfriendly chemicals are used in the processing industry. These chemicals require careful monitoring during use, transportation, and handling. In a basic text it is not possible to cover the above-mentioned sensors' availability and precautions in their applications, but just to make the student aware of their existence. Analysis labs and control rooms must meet code; further information can be obtained from the ISA series RP 60 practices. All processing plants and labs will have an alarm system which can shut down certain operations if a problem occurs. These systems are regularly tested and are often duplicated to provide built-in fail-safe features such as redundancy as protection against sensor failure.

10.5 Sound and Light

10.5.1 Sound and light formulas

The measurement of sound and light is important as it relates to the sense of hearing and sight, as well as many industrial applications such as the use of sound waves for the detection of flaws in solids and in location and linear distance measurement. Sound pressure waves can induce mechanical vibration and hence failure. Excessive sound levels produce noise pollution. Light and its measurement is used in many industrial applications for high-accuracy linear measurements, location of overheating (infrared), object location and position measurements, photo processing, scanning, readers (bar codes), and so forth.

Sound waves are pressure waves that travel through air, gas, solids, and liquids, but cannot travel through space or a vacuum unlike radio (electromagnetic) waves. Pressure waves can have frequencies up to about 50 kHz. Sound or sonic waves start at 16 Hz and go up to 20 kHz; above 30 kHz sound waves are ultrasonic. Sound waves travel through air at about 340 m/s (depends on temperature, pressure, and the like.). The amplitude or loudness of sound is measured in phons.

Sound pressure levels (SPL) are units often used in the measurement of sound levels and are defined as the difference in pressure between the maximum pressure at a point and the average pressure at that point. The units of pressure are normally expressed as follows:

$$1 \text{ dyn/cm}^2 = 1 \text{ ubar} = 1.45 \times 10^{-5} \text{ psi} \tag{10.8}$$

where $1 \text{ N} = 10^5 \text{ dyn}$.

Decibel (dB) is a logarithmic measure used to measure and compare amplitudes and power levels in electrical units, sound, light, and the like. The sensitivity of the ears and eyes are logarithmic. To compare different sound intensities the following applies:

$$\text{Sound level ratio in dB} = 10 \log_{10}\left(\frac{I_1}{I_2}\right) \tag{10.9}$$

where I_1 and I_2 are the sound intensities at two different locations and are scalar units. A reference level (for I_2) is 10^{-16} W/cm^2 (average level of sound that can be detected by the human ear at 1 kHz) to measure sound levels.

When comparing different pressure levels the following is used:

$$\text{Pressure level ratio in dB} = 20 \log_{10}\left(\frac{P_1}{P_2}\right) \tag{10.10}$$

where P_1 and P_2 are the pressures at two different locations (note pressure is a measure of sound power, hence 20 log). For P_2, 20 µN/m^2 is accepted as the average pressure level of sound that can be detected by the human ear at 1 kHz and is therefore, the reference level for measuring sound pressures.

Typical figures for SPL are as follows:

Threshold of pain	140 to 150 dB
Rocket engines	170 to 180 dB
Factory	80 to 100 dB

Light is ultra-high frequency electromagnetic wave that travels at 3×10^8 m/s. Light amplitude is measured in foot-candles (fc) or lux (lx). The wavelength of visible light is from 4 to 7×10^{-7} m. Longer wavelengths of electromagnetic waves are termed infrared and shorter wavelengths, ultraviolet. Light wavelengths are sometimes expressed in terms of angstroms (Å) where $1 \text{ Å} = 1 \times 10^{-10}$ m.

Example 10.6 What is the wavelength of light in Å, if the wavelength in meters is 500 nm?

$$1 \text{ Å} = 10^{-10} \text{ m}$$
$$500 \text{ nm} = 500 \times 10^{-9}/10^{-10} \text{ Å} = 5000 \text{ Å}$$

Intensity is the brightness of light. The unit of measurement of light intensity in the English system is the foot-candle (fc), which is one lumen per square foot (lm/ft^2). In the SI system the unit is the lux (lx) which is one lumen per square meter (lm/m^2). The phot (ph) is also used and is defined as one lumen per square centimeter (lm/cm^2). The lumen replaces the candela (cd) in the SI system. The dB is also used for the comparison of light intensity as follows:

$$\text{Light level ratio in dB} = 10 \log_{10}\left(\frac{\Phi_1}{\Phi_2}\right) \quad (10.11)$$

where Φ_1 and Φ_2 are the light intensity at two different points.

The change in intensity levels for both sound and light from a source is given by the following equation:

$$\text{Change in levels} = 10 \log_{10}\left(\frac{d_1}{d_2}\right) \quad (10.12)$$

where d_1 and d_2 are the distances from the source to the points being considered.

Example 10.7 Two points are 65 and 84 ft from a light bulb. What is the difference in the light intensity at the two points?

$$\text{Difference} = 10 \log_{10}\left(\frac{65}{84}\right) = -1.11 \text{ dB}$$

X-rays should be mentioned at this point as they are used in the process control industry and are also electromagnetic waves. X-rays are used primarily as inspection tools; the rays can be sensed by some light-sensing cells and can be very hazardous if proper precautions are not taken.

10.5.2 Sound and light measuring devices

Microphones are transducers used to convert sound levels into electrical signals, i.e., electromagnetic, capacitance, ribbon, crystal, carbon, and piezoelectric microphones can be used. The electrical signals can then be analyzed in a spectrum analyzer for the various frequencies contained in the sounds or just to measure amplitude.

Sound level meter is the term given to any of the variety of meters for measuring and analyzing sounds.

Photocells are used for the detection and conversion of light intensity into electrical signals. Photocells can be classified as photovoltaic, photoconductive, photoemissive, and semiconductor.

Photovoltaic cells develop an emf in the presence of light. Copper oxide and selenium are examples of photovoltaic materials. A microammeter calibrated in lux (lm/m^2) is connected across the cells and measures the current output.

Photoconductive devices change their resistance with light intensity. Such materials are selenium, zirconium oxide, aluminum oxide, and cadmium sulfide.

Photoemissive materials, such as mixtures of rare earth elements (cesium oxide), liberate electrons in the presence of light.

Semiconductors are photosensitive and are commercially available as photodiodes and phototransistors. Light generates hole-electron pairs, which causes leakage in reversed biased diodes and base current in phototransistors. Commercial high-resolution optical sensors are available with the electronics integrated onto a single die to give temperature compensation and a linear voltage output with incident light intensity are also commercially available. Such a device is the TSL 250. Also commercially available are infrared (IR) light-to-voltage converters (TSL 260) and light-to-frequency converters (TSL 230). Note, the TSL family is manufactured by Texas Instruments.

10.5.3 Light sources

Incandescent light is produced by electrically heating a resistive filament or the burning of certain combustible materials. A large portion of the energy emitted is in the infrared spectrum as well as the visible spectrum.

Atomic-type sources cover gas discharge devices such as neon and fluorescent lights.

Laser emissions are obtained by excitation of the atoms of certain elements.

Semiconductor diodes (LED) are the most common commercially available light sources used in industry. When forward biased, the diodes emit light in the visible or IR region. Certain semiconductor diodes emit a narrow band of wavelength of visible rays; the color is determined by material and doping. A list of LEDs and their color is given in Table 10.1.

10.5.4 Sound and light application considerations

Selection of sensors for the measurement of sound and light intensity will depend on the application. In instrumentation a uniform sensitivity over a wide frequency range requires low inherent noise levels, consistent sensitivity

TABLE 10.1 LED Characteristics

Material	Dopant	Wavelength (nm)	Color
GaAs	Zn	900	IR
GaP	Zn	700	Red
GaAsP	N	632	Orange
GaAsP	N	589	Yellow
GaP	N	570	Green
SiC	—	490	Blue

Photoresistor   Photodiode

Phototransistor   Photovoltaic (Solar) Cell

Figure 10.7 Schematic symbols for opto sensors.

with life, and a means of screening out unwanted noise and light from other sources.

In some applications, such as the sensing of an optical disc, it is only necessary to detect absence or presence of a signal, which enables the use of cheap and simple sensors. For light detection, the phototransistor is being used very widely, because of the ability with integrated circuits to put temperature correction and amplification in the same package for high-sensitivity. The device is cost effective and has good longevity.

Figure 10.7 shows the schematic symbols used for optoelectronic sensors and Table 10.2 gives a comparison of photosensor characteristics.

TABLE 10.2 Summary of Opto Sensor Characteristics

Type	Device	Response (μm)	Response Time	Advantages	Disadvantages
Photo-conductive	Photo-resistor CdS	0.6–0.9	100 ms	Small, high-sensitivity, low cost	Slow, hystersis, limited temperature
	Photo-resistor CdSe	0.6–0.9	10 ms	Small, high-sensitivity, low cost	Slow, hystersis, limited temperature
Semi-conductor	Photo-diode	0.4–0.9	1 ns	Very fast, good linearity, low noise	Low-level output
	Phototransistor	0.25–1.1	1 μs		Low frequency response, nonlinear
Photo-voltaic	Solar cell	0.35–0.75	20 μs	Linear, self powered	Slow, low-level output

Summary

A number of different types of sensors were discussed in this chapter. Sensors for measuring position, speed, and acceleration were introduced. The concepts of force, torque, and load measurements were discussed together with measuring devices. Also covered in this chapter were smoke and chemical sensors and an introduction to sound and light measurements and instruments.

The salient points covered are as follows:

1. The basic terms and standards used in linear and rotational measurements and the sensors used for the measurement of absolute and incremental position, velocity, and acceleration
2. Optical and magnetic sensors and their use as position measuring devices in linear as well as rotational applications and sensors used for distance measurement
3. Definitions of force, torque, couples, and load and the use of mechanical forces in weight measurements
4. Stress in materials, the use of strain gauges for its measurement, and instruments used for measurement
5. Smoke and chemical sensors were introduced with sensors and applications
6. An introduction to sound and light units and the units used in their measurement with examples
7. Sound and optical sensing devices are given and the type of semiconductor devices used for light generation with their color spectrum

Problems

10.1 What force is necessary to accelerate a mass of 17 lb at 21 ft/s^2?

10.2 What is the acceleration of 81 kg, when acted upon by a force of 55 N?

10.3 What torque does a force of 33 lb produce 13 ft perpendicular from a fulcrum?

10.4 What force is required to produce a torque of 11 N·m, if the force is 13 m from the fulcrum?

10.5 A couple of 53 N·m is produced by two equal forces of 15 N, how far are they apart?

10.6 A couple of 38 lb-ft is produced by two equal forces 8 ft apart. What is the magnitude of the forces?

10.7 A balance has a reference weight of 10 kg, 0.5 m from the fulcrum. How far from the fulcrum must a weight of 16 kg be placed to be balanced?

10.8 A reference weight of 15 lb is placed 2 ft from the fulcrum to balance a weight 4.7 ft from the fulcrum. What is the weight?

10.9 A spring balance has a spring constant of 3 lb/in, a basket with 6-lb potatoes extends the spring 2.7 in. What is the weight of the basket?

10.10 A spring balance has a spring constant of 2.3 N/m. What is the deflection of the balance when it is loaded with 0.73 N?

10.11 A force of 10 N is applied to a piston 150 cm in diameter in a closed cylinder. What is the pressure in the cylinder?

10.12 What force is applied to a 9.2-in piston in a cylinder, if the pressure in the cylinder is 23 psi?

10.13 What is the wavelength of 13 kHz sound waves?

10.14 What is the frequency of radar waves whose wavelength is 2.5 cm?

10.15 A person is 375 m from a bell and a second person is 125 m from the bell. What is the difference in the sound levels of the bell rings heard by the two people?

10.16 What is the sound pressure level corresponding to 67 dB?

10.17 The light intensity 20 ft from a light bulb is 3.83 dB higher than at a second point. What is the distance of the second point from the bulb?

10.18 What is the change in light intensity when the distance from the light source is increased from 35 to 85 ft?

10.19 What is the angular displacement that can be sensed by an angular displacement sensor, if the circular disc on a shaft has 115 slots and the photodiode array has 16 diodes? See Fig. 10.3.

10.20 The rotational speed of a steel cog wheel is being sensed with a magnetic sensor. If the wheel has 63 teeth and is rotating at 1021 r/min, what is the frequency of the output pulses?

Chapter 11

Actuators and Control

Chapter Objectives

This chapter is an introduction to actuators and actuator control and will help you understand the operation of control devices and their use in flow control.
Topics covered in this chapter are as follows:

- The operation and use of various types of self-regulating devices and loading used in gas regulators
- Closed loop pneumatic, hydraulic, and electrical liquid flow valves
- Fail-safe operation in valves
- The various types of valves in use
- Valve characteristics
- Electronic power control devices
- Methods of applying feedback for position control
- Relays and motors for actuator control

This section deals with control devices used for regulating temperature, pressure, controlling liquid, and gas flow in industrial processing. The devices can be self-regulating or under the control of a central processing system that can be monitoring and controlling many variables.

11.1 Introduction

This section will discuss actuators and regulators and their use for the control of gas flow, liquid flow, and pressure control. In many processes this involves the control of many thousands of cubic meters of a liquid or the control of large forces, as would be the case in a steel rolling mill from low-level analog, digital, or pneumatic signals. Temperature can also normally be controlled by regulating gas and/or liquid flow. Control loops can be local self-regulating loops under pneumatic, hydraulic, or electrical control, or the loops can be processor controlled with additional position feedback loops. Electrical signals from a controller are low-level signals that require the use of relays for power control or amplification and power-switching devices, and possibly opto-isolators for isolation. These power control devices are normally at the point of use so that electrically controlled actuators and motors can be supplied directly from the power lines.

11.2 Pressure Controllers

11.2.1 Regulators

Gases used in industrial processing, such as oxygen, nitrogen, hydrogen, and propane, are stored in high-pressure containers in liquid form. The high-pressure gases from above the liquid are reduced in pressure and regulated with gas regulators to a lower pounds per square inch before they can be distributed through the facility. The gas lines may have additional regulators at the point of use.

A *spring-controlled regulator* is an internally controlled pressure regulator and is shown in Fig. 11.1a. Initially, the spring holds the inlet valve open and gas under pressure flows into the main cylinder at a rate higher than it can exit the cylinder. As the pressure in the cylinder increases, a predetermined pressure is reached where the spring loaded diaphragm starts to move up, causing the valve to partially close, i.e., the pressure on the diaphragm controls the flow of gas into the cylinder to maintain a constant pressure in the main cylinder and at the output, regardless of the flow rate (ideally). The output pressure can be adjusted by the spring screw adjustment.

A *weight-controlled regulator* is shown in Fig. 11.1b. The internally controlled regulator has a weight-loaded diaphragm. The operation is the same as the spring-loaded diaphragm except the spring is replaced with a weight. The pressure can be adjusted by the position of a sliding weight on a cantilever arm.

A *pressure-controlled diaphragm regulator* is shown in Fig. 11.2a. The internally controlled regulator has a pressure-loaded diaphragm. Pressure from a

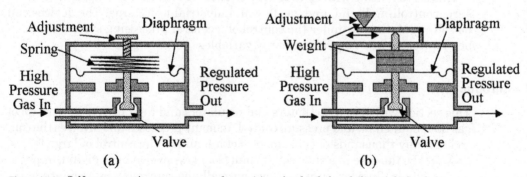

Figure 11.1 Self-compensating pressure regulators (*a*) spring loaded and (*b*) weight loaded.

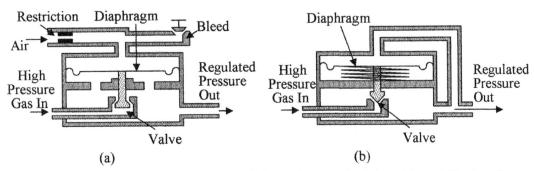

Figure 11.2 Self-compensating pressure regulators (a) internal pressure-loaded regulator and (b) externally connected spring-loaded regulator.

regulated external air or gas supply is used to load the diaphragm via a restriction. The pressure to the regulator can then be adjusted by an adjustable bleed valve, which in turn is used to set the output pressure of the regulator.

Externally connected spring diaphragm regulator is shown in Fig. 11.2b. The cross section shows an externally connected spring-loaded pressure regulator. The spring holds the valve open until the output pressure, which is fed to the upper surface of the diaphragm, overcomes the force of the spring on the diaphragm, and starts to close the valve, hence regulating the output pressure. Note that the valve is inverted from the internal regulator and the internal pressure is isolated from the lower side of the diaphragm. Weight- and air-loaded diaphragms are also available for externally connected regulators.

Pilot-operated pressure regulators can use an internal or external pilot for feedback signal amplification and control. The pilot is a small regulator positioned between the pressure connection to the regulator and the loading pressure on the diaphragm. Figure 11.3a shows such an externally connected pilot regulator. The pressure from the output of the regulator is used to control the pilot, which

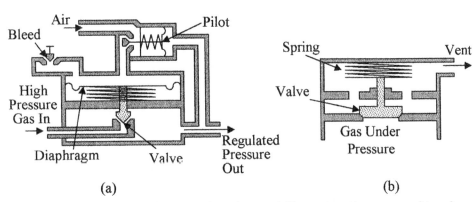

Figure 11.3 Shown are (a) a pilot-operated regulator and (b) an automatic pressure safety valve.

in turn amplifies the signal and controls the pressure from the air supply to the diaphragm, giving greater control than that available with the internal pressure control diaphragm. A slight change in the output pressure is required to produce a full pressure range change of the regulator giving a high gain system for good output pressure regulation.

An *Instrument pilot-operated pressure regulator* is similar to the pilot-operated pressure regulator but has a proportional band adjustment included, giving a gain or sensitivity control feature to provide greater flexibility in control.

11.2.2 Safety valves

Safety valves are fitted to all high-pressure containers from steam generators to domestic water heaters (see Fig. 11.3b). The valve is closed until the pressure on the lower face of the valve reaches a predetermined level set by the spring. When this level is reached, the valve moves up allowing the excess pressure to escape through the vent.

11.2.3 Level regulators

Level regulators are in common use in industry to maintain a constant fluid pressure, or a constant fluid supply to a process, or in waste storage. Level regulators can be a simple float and valve arrangement as shown in Fig. 11.4a to using capacitive sensors as given in Chap. 6 to control a remote pump. The arrangement shown in Fig. 11.4a is used to control water levels in many applications. When the fluid level drops due to use, the float moves downward opening the inlet valve and allowing fluid to flow into the tank. As the tank fills, the float rises, causing the inlet valve to close, thus maintaining a constant level and preventing the tank from overflowing.

Figure 11.4b shows an example of a self-emptying reservoir when a predetermined fluid level is reached, as may be used in a waste holding tank. As the tank fills, the float rises to where the connecting link from the float to the valve

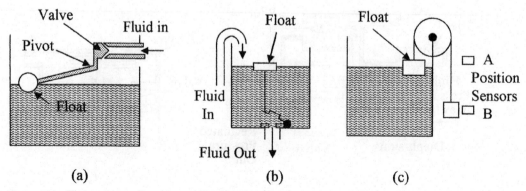

Figure 11.4 Various types of regulators are shown (a) automatic fluid level controller, (b) automatic emptying of a storage tank when full, and (c) means of detecting full level or empty level in a fluid reservoir.

becomes taut and overcomes the hydrostatic pressure on and lifting the outlet valve. Once lifted, the fluid pressure under the valve balances the pressure above the valve and the buoyancy of the valve will keep it open until the tank is empty, then it will close. Once closed the reservoir will start to fill and the fluid pressure on the top surface of the valve will hold it closed. The automatic fluid leveler in Fig. 11.4a can be combined with the emptying system in Fig. 11.4b. In this case, the outlet valve is manually or automatically operated to deliver a known volume of liquid to a process, as required. The container automatically refills for the next operation cycle.

The position of the weight in Fig. 11.4c is controlled by the float. The position of the weight is monitored by position sensors A and B. When the weight is in position A (container empty), the sensor can be used to turn on a pump to fill the tank and when sensor B senses the weight (container full) it can be used to turn the pump off. The weight can be made of a magnetic material and the level sensors would be Hall effect or *magneto resistive element* (MRE) devices.

11.3 Flow Control Actuators

When a change in a measured variable with respect to a reference has been sensed, it is necessary to apply a control signal to an actuator to make corrections to an input controlled variable to bring the measured variable back to its preset value. In most cases any change in the variables, i.e., temperature, pressure, mixing ingredients, and level, can be corrected by controlling flow rates. Hence, actuators are in general used for flow rate control and can be electrically, pneumatically, or hydraulically controlled. Actuators can be self-operating in local feedback loops in such applications as temperature sensing with direct hydraulic or pneumatic valve control, pressure regulators, and float level controllers. There are two common types of variable aperture actuators used for flow control; they are the globe valve and the butterfly valve.

11.3.1 Globe valve

The globe valve's cross section is shown in Fig. 11.5a. The actuator can be driven electrically using a solenoid or motor, pneumatically or hydraulically. The actuator determines the speed of travel and distance the valve shaft travels. The globe-type valve can be designed for quick opening, linear, or equal percentage operation. In equal percentage operation the flow is proportional to the percentage the valve is open, or there is a log relationship between the flow and valve travel. The shape of the plug determines the flow characteristics of the actuator and is normally described in terms of percentage of flow versus percentage of lift or travel.

The valve plug shown in Fig. 11.5a gives a linear relationship between flow and lift. The characteristic is given in Fig. 11.5b. Also shown in the graph are the characteristics for a quick opening plug and an equal percentage plug to illustrate some of the characteristics that can be obtained from the large number of plugs that are available. The selection of the type of control plug should be carefully chosen for any particular application. The type will depend on a careful analysis of the process characteristics, i.e., if the load changes are linear a

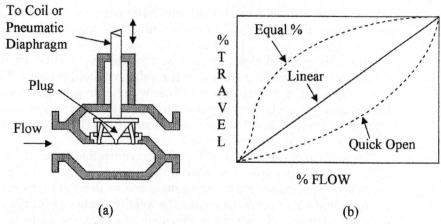

Figure 11.5 Cross section of (*a*) a globe valve with a linear flow control plug and (*b*) different flow patterns for various plugs versus plug travel.

linear plug should be used. Conversely, if the load changes are nonlinear a plug with the appropriate nonlinear characteristic should be used.

The globe valve can be straight through with single seating as illustrated in Fig. 11.5*a* or can be configured with double seating, which is used to reduce the actuator operating force, but is expensive, difficult to adjust and maintain, and does not have a tight seal when shutoff. Angle valves are also available, i.e., the output port is at right angles or 45° to the input port.

Many other configurations of the globe valve are available. Illustrated in Fig. 11.6*a* is a two-way valve (diverging type), which is used to switch the

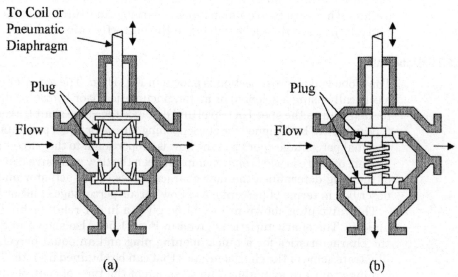

Figure 11.6 Cross sections of globe valve configurations: (*a*) two-way valve and (*b*) three-position valve.

incoming flow from one exit to another. When the valve stem is up the lower port is closed and the incoming liquid exits to the right, and when the valve is down the upper port is closed and the liquid exits from the bottom. Also available is a converging-type valve, which is used to switch either of the two incoming flows to a single output. Figure 11.6b illustrates a three-way valve. In the neutral position both exit ports are held closed by the spring. When the valve stem moves down the top port is opened and when the valve stem moves up from the neutral position the lower port is opened.

Other types of globe valves are the needle valve (less than 1-in diameter), the balanced cage-guided valve, and the split body valve. In the cage-guided valve, the plug is grooved to balance the pressure in the valve body. The valve has good sealing when shut off. The split body valve is designed for ease of maintenance and can be more cost effective than the standard globe valve, but pipe stresses can be transmitted to the valve and cause it to leak. Globe valves are not well suited for use with slurries.

11.3.2 Butterfly valve

The butterfly valve is shown in Fig. 11.7a and its flow versus travel characteristics are shown in Fig. 11.7b. The relation between flow and lift is approximately equal percentage up to about 50 percent open, after which it is linear. Butterfly valves offer high capacity at low cost, are simple in design, easy to install, and have tight closure. The torsion force on the shaft increases until open up to 70° and then reverses.

11.3.3 Other valve types

A number of other types of valves are in common use. They are the weir diaphragm, ball, and rotary plug valves. The cross sections of these valves are shown in Fig. 11.8.

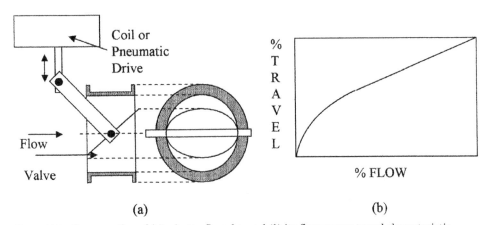

Figure 11.7 Cross section of (a) a butterfly valve and (b) its flow versus travel characteristic.

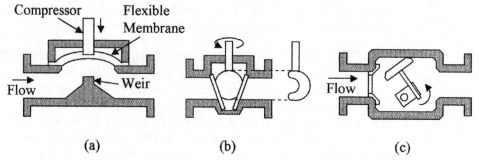

Figure 11.8 Different valve types (*a*) diaphragm, (*b*) one-piece ball valve, and (*c*) rotary plug valve.

A *weir-type diaphragm valve* is shown in Fig. 11.8*a*. The valve is shown open; closure is achieved by forcing a flexible membrane down onto the weir. Diaphragm valves are good for slurries and liquids with suspended solids, are low cost devices, but tend to require high maintenance, and have poor flow characteristics.

A one-piece *ball valve* is shown in Fig. 11.8*b*. The valve is a partial sphere that rotates. The valve tends to be slow to open. Other than the one shown in the figure, the ball valve is available in other configurations also with spheres of various shapes for different flow characteristics. The valve is good for slurries and liquids with solid matter because of its self-cleaning operation. Ball valves have tight turnoff characteristics, are simple in design, and have greater capacity than similar-sized globe valves.

An eccentric *rotary plug valve* is shown in Fig. 11.8*c*. The valve is of medium cost but requires less closing force than many valves and can be used for forward or reverse flow. The valve has tight shutoff with positive seating action, has high capacity, and can be used with corrosive liquids.

11.3.4 Valve characteristics

Other factors that determine the choice of valve type are corrosion resistance, operating temperature ranges, high and low pressures, velocities, and fluids containing solids. Correct valve installation is essential; vendor recommendations must be carefully followed. In situations where sludge or solid particulates can be trapped upstream of a valve, a means of purging the pipe must be available. To minimize disturbances and obtain good flow characteristics a clear run of 1 to 5 pipe diameters up and down stream should be allowed.

Valve sizing is based on pressure loss. Valves are given a C_V number that is based on test results. The C_V number is the number of gallons of water flowing per minute through a fully open valve at 60°F (15.5°C) that will cause a pressure drop of 1 psi (6.9 kPa). It implies that when flowing through the fully opened valve, it will have a pressure drop of 1 psi (6.9 kPa), i.e., a valve with a C_V of 25 will have a pressure drop of 1 psi when 25 gal of water per minute is flowing

TABLE 11.1 Valve Characteristics

Parameter	Globe	Diaphragm	Ball	Butterfly	Rotary plug
Size	1 to 36 in	1 to 20 in	1 to 24 in	2 to 36 in	1 to 12 in
Slurries	No	Yes	Yes	No	Yes
Temperature range (°C)	−200 to 540	−40 to 150	−200 to 400	−50 to 250	−200 to 400
Quick-opening	Yes	Yes	No	No	No
Linear	Yes	No	Yes	No	Yes
Equal %	Yes	No	Yes	Yes	Yes
Control range	20:1 to 100:1	3:1 to 15:1	50:1 to 350:1	15:1 to 50:1	30:1 to 100:1
Capacity (C_V) (d = Dia.)	10 to 12 × d^2	14 to 22 × d^2	14 to 24 × d^2	12 to 35 × d^2	12 to 14 × d^2

through it. For liquids, the relation between pressure drop P_d (pounds per square inch), flow rate Q (gallon per minute), and C_V is given by

$$C_V = Q \times \sqrt{(SG/P_d)} \tag{11.1}$$

where SG is the specific gravity of the liquid.

Example 11.1 What is the C_V of a valve, if there is a pressure drop of 3.5 psi when 2.3 gal per second of a liquid with a specific weight (SW) of 60 lb/ft^3 is flowing?

$$C_V = 2.3 \times 60 \sqrt{\frac{60}{62.4 \times 3.5}} = 138 \times 0.52 = 72.3$$

Table 11.1 gives a comparison of some of the valve characteristics; the values shown are typical of the devices available and may be exceeded by some manufacturers with new designs and materials.

11.3.5 Valve fail safe

An important consideration in many systems is the position of the actuators when there is a loss of power, i.e., will chemicals or the fuel to the heaters continue to flow or will a total system shut down occur? Figure 11.9 shows an example of a pneumatically or hydraulically operated globe valve design that can be configured to go to the open or closed position during a system failure. The modes of failure are determined by simply changing the spring position and the pressure port.

In Fig. 11.9a the globe valve is closed by applying pressure to the pressure port to oppose the spring action. If the system fails, i.e., if there is a loss of pneumatic pressure, the spring acting on the piston will force the valve to revert back to its open position. In Fig. 11.9b the spring is removed from below the piston to a position above the piston and the inlet and exhaust ports are reversed. In this case the valve is opened by the applied pressure working against the spring action. If the system fails and there is a loss of control pressure, the spring action

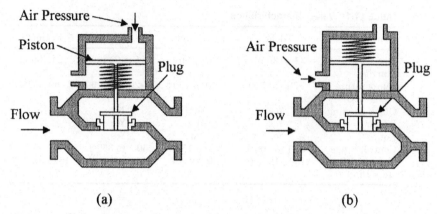

Figure 11.9 Fail-safe pneumatic or hydraulic operated valves: with the loss of operating pressure on the valve, the valve in (a) will open and the valve in (b) will close.

will force the piston down and close the valve. Similar fail-safe electrically and hydraulically operated valves are available. Two-way and three-way fail-safe valves are also available which can be configured to be in a specific position when the operating system fails.

11.4 Power Control

Electrical power for actuator operation can be controlled from low-level analog and digital signals using electronic power devices or magnetic contactors. Magnetic contactors have a lower ON resistance than electronic devices but require higher drive power. Contactors provide voltage isolation between the control signals and output power circuits, but are slow to switch, have lower current handling capability than electronic power devices, and have a limited switching life. In electronic devices the problem of electrical isolation between drive circuits and output power circuits can easily be overcome with the use of opto-isolators. Electronic power devices have excellent longevity and are very advantageous due to their switching speeds in variable power control circuits.

11.4.1 Electronic devices

A number of electronic devices such as the silicon-controlled rectifier (SCR), TRIAC, and metal-oxide semiconductor (MOS) devices can be used to control several hundred kilowatts of power from low-level electrical signals. Electronic power control devices fall into two categories. First, triggered devices such as SCR and TRIAC that are triggered by a pulse on the gate into the conduction state, once triggered can only be turned off by reducing the anode/cathode current to below their sustaining current, i.e., when the supply voltage/current drops to zero. But these devices can block high reverse voltages. Hence, they can be extensively used in ac circuits where the supply regularly transcends through zero turning the device OFF automatically. The second group of devices are Darlington bipolar junction transistors (BJT), power metal-oxide semiconductor field effect transistor (MOSFET), insulated gate bipolar transistors (IGBT), and MOS-controlled thyristors (MCT). These devices are turned ON and OFF by an input control signal, but do not have the capability of high reverse voltage blocking. Hence, this group of devices are more commonly used with dc power supplies or biased to prevent a reverse voltage across the device.

The SCR is a current-operated device and can only be triggered to conduct in one direction, i.e., when used with an ac supply it blocks the negative half-cycle and will only conduct on the positive half cycle, when triggered. Once triggered, the SCR remains ON for the remaining portion of the half-cycle. Figure 11.10a shows the circuit of an SCR with a load. Figures 11.10b and c show the effects of triggering on the load voltage (V_L). By varying the triggering in relation to the positive half cycle, the power in the load can be controlled from 0 to 50 percent of the total available power. Power can be controlled from 50 to 100 percent by putting a diode in parallel with the SCR to conduct current on the negative half cycle. Light activated SCRs are also available.

One method of triggering the SCR is shown in Fig. 11.11a with the corresponding circuit waveforms shown in Fig. 11.11b. During the positive half-cycle the capacitor C is charged via R_1 and R_2 until the triggering point of the SCR is reached. The diode can be connected on either side of the load. The advantage of connecting the diode to the SCR side of the load is to turn OFF the voltage to the gate when the SCR is fired, thus, reducing dissipation. The diode is used to block the negative half-cycle from putting a high negative voltage on the gate and damaging the SCR. The zener diode is used to clamp the positive going

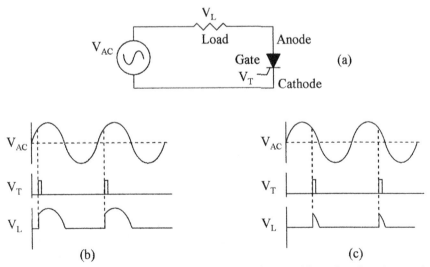

Figure 11.10 (*a*) SCR circuit with load, (*b*) Waveforms with early triggering, and (*c*) Waveforms with late (low power to load) triggering.

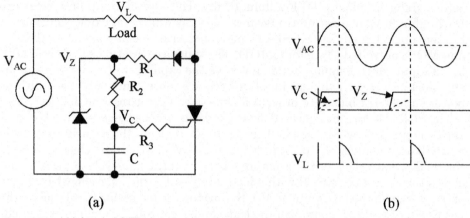

Figure 11.11 (a) A typical SCR triggering circuit with trigger point control and (b) triggering waveforms.

half-cycle at a fixed voltage (V_Z) so that the capacitor (V_C) has a fixed aiming voltage, giving a linear relation between triggering time and potentiometer setting. This is shown by V_Z and V_C in Fig. 11.11b.

Example 11.2 In Fig. 11.11 a SCR with a 5 V gate trigger level is used with a 12 V zener diode; the capacitor is 0.15 µF. What is the value of R_2 to give full control of the power to the load down to zero?

Time duration of half sine wave at 60 Hz = 1/60 × 2 = 8.3 ms
Charging time can be found from capacitor charging equation $V_C = V_0 (1 - e^{-t/RC})$

$$5 = 12(1 - e^{-t/RC})$$

from which $t = 0.54$ RC = 8.3 ms

$$R = 8.3 \times 10^6/0.54 \times 0.15 \times 10^3 = 102.5 \text{ k}\Omega$$

A control of 0 to 100 percent can be obtained with a single SCR in a bridge circuit as shown in Fig. 11.12a; the waveforms are shown in Fig. 11.12b. The bridge circuit changes the negative going half-cycles into positive half-cycles so that the SCR only sees positive half-cycles and is triggered during every half-cycle, and is turned OFF every half-cycle when the supply voltage goes to zero. As shown in the Fig.11.12a the system is controlled by a low level signal coupled to the SCR trigger circuit via an opto-isolator. The triggering point is set by potentiometer R and capacitor C; as the SCR only sees positive voltages, the diode in not required. For cheapness the zener diode is omitted. As in the previous figure, resistor R can be connected to either side of the load.

The DIAC is a semiconductor device developed for trigger control primarily for use with TRIACs. Figure 11.13a shows the symbol for the device and (b) the device's characteristic. The DIAC is a two-terminal symmetrical switching device. As the voltage increases across the device, little current flows until the

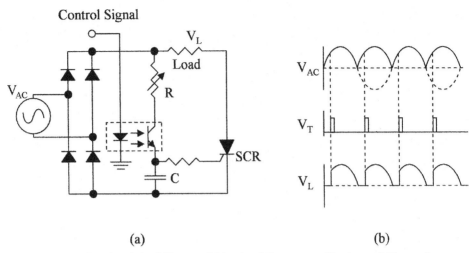

Figure 11.12 Bridge circuit for SCR control (a) using full-wave rectification and (b) waveforms.

breakdown voltage V_L is reached. At this point the device breaks down and conducts as shown. The breakdown occurs with both positive and negative voltages. The breakdown voltage of the DIAC is used to set the trigger voltage for the TRIAC; when the device breaks down the TRIAC triggers.

TRIACs can be considered as two reversed SCRs connected in parallel. They can be triggered on both the positive and negative half-cycles of the ac waveform. A circuit for triggering a TRIAC is shown in Fig. 11.14a with the associated waveforms shown in Fig. 11.14b. The TRIAC can be used to control power to the load from 0 to 100 percent by controlling the trigger points with respect to the ac sine wave. As the ac voltage increases from zero, V_Z is clamped by the zener diodes in both the positive and negative directions. The capacitor C is then charged via R_2 until the breakdown voltage of the DIAC is reached and the TRIAC is triggered on both the positive and negative half-cycles as shown by the waveforms in Fig. 11.14b.

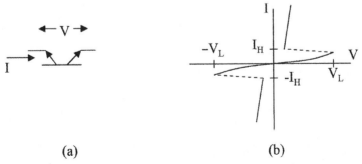

Figure 11.13 DIAC used in SCR and TRIAC triggering circuits (a) symbol and (b) characteristic.

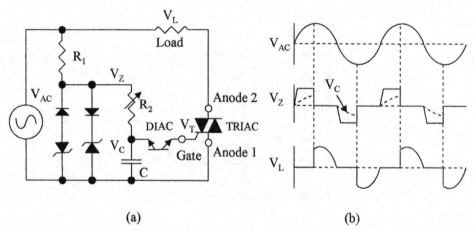

Figure 11.14 A TRIAC can control power from 0 to 100 percent (a) shows the TRIAC power control circuit and (b) the circuit waveforms.

Example 11.3 A TRIAC is used to supply 750 amps to a load from 120 V supply. What is the maximum power that can be supplied to the load and the power loss in the TRIAC? Assume the voltage drop across the TRIAC is 2.1 V.

$$\text{Power loss in TRIAC} = 2.1 \times 750 \text{ W} = 1.575 \text{ kW}$$

$$\text{Power from supply} = 750 \times 120 \text{ W} = 90 \text{ kW}$$

$$\text{Power to load} = 90 - 1.575 \text{ kW} = 88.425 \text{ kW}$$

This example illustrates that the efficiency of the switch >98 percent, and also the high dissipation that can occur in the switch and the need for cooling fins with low thermal resistance. Precautions in the design of power switching circuits, choices of devices for specific applications, and thermal limitations are outside the scope of this book. Device data sheets must be consulted and advice obtained from device manufacturers before designing power controllers.

Power devices that have an input control are as follows:

1. *Darlington Bipolar Junction Transistors (BJT)* are current-controlled devices. Power bipolar devices have low gain and so are normally used in a Darlington configuration to give high current gain and the ability to control high currents with low drive currents.

2. *Power MOSFETs* are voltage-controlled devices designed for high-speed operation, but their high saturation voltage and temperature sensitivity limits their application in power circuits.

3. *Insulated Gate Bipolar Transistor (IGBT)*, as opposed to the Darlington bipolar configuration, is controlled by a MOS transistor making it a voltage-controlled device. The IGBT has fast switching times. Older devices had a high saturation voltage; newer devices have a saturation voltage about the same as a BJT.

4. *MOS-Controlled Thyristor (MCT)* is a voltage-controlled device with a low saturation voltage and medium speed switching characteristics.

TABLE 11.2 Comparison of Power Device Characteristics

Device	Power handling	Saturation (Volt)	Turnon time	Turnoff time
SCR	2 kV 1.5 kA	1.6 V	20 µs	N/A
TRIAC	2 kV 1 kA	2.1 V	20 µs	N/A
BJT	1.2 kV 800 A	1.9 V	2 µs	5 µs
MOSFET	500 V 50 A	3.2 V	90 ns	140 ns
IGBT	1.2 kV 800 A	1.9 V	0.9 µs	200 ns
MCT	600 V 60 A	1.1 V	1.0 µs	2.1 µs

A comparison of the power devices characteristics is given in Table 11.2. These devices are used for power and motor control. Applications include rectification of multiphase ac power to give a variable voltage dc power level output or the control of dc motors from an ac power source, the control of multiphase motors from a dc power source, or the conversion of dc power to multiphase ac power.

11.4.2 Magnetic control devices

A signal from a controller is a low-level signal but can be amplified to control an actuator or small motor. Power for actuators are normally generated close to the point of use to prevent energy loss in the leads and to prevent large currents from flowing in the ground return lines to the controller to minimize offset and ground line noise. In Fig. 11.15a a power transistor is used to drive a solenoid valve. A diode

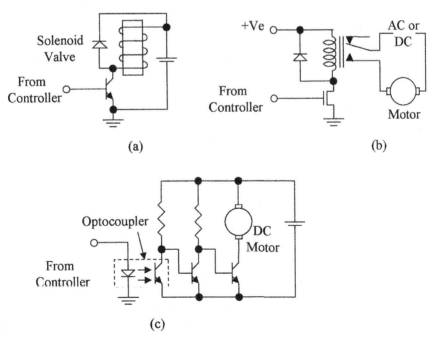

Figure 11.15 Options are shown for driving a motor or actuator from a controller output (a) via a power transistor, (b) via a relay, and (c) using an opto-coupler.

is used across the solenoid to protect the transistor from the high voltage inductive overshoot that occurs on switch-off. In Fig. 11.15b a MOS device is used to drive a motor control relay. Because of the isolation the relay gives between the driving circuit and the motor circuit, the motor and power supply can be either dc or ac. Such a circuit can be extended to driving multicontact relays to control three-phase ac motors and multiple signal paths. Relays for switching high currents and voltages that are used for motor control are called contactors.

Figure 11.15c shows the use of an opto-coupler to isolate the controller from the motor circuit. While both circuits are electrically isolated, the circuit as shown can only be used to drive a dc motor. However, because of the isolation given by the opto-coupler, the circuit can be expanded to drive three-phase ac motors. The opto-coupler consists of a light emitting diode (LED) optically coupled to a phototransistor; a current (10 to 30 mA) activates the diode; light from the diode turns ON the phototransistor. When there is no current flowing in the LED, no light is emitted and the phototransistor is OFF. As previously mentioned, solid state relays are available that have the power device (TRIAC) included in the package with the opto-coupler for direct motor control.

Contactors are designed for switching high currents and voltages, such as are used in motor control applications. A single-pole single-throw double-break contactor is shown in Fig. 11.16. In Fig. 11.16a the contactor is shown de-energized and the contacts are open. When a current is passed through the coil, the magnetic field in the core attracts and pulls in the soft iron keeper which closes the contacts as shown in Fig. 11.16b. Contactors can have multiple contacts for multiphase motors. Contact material is critical, as chemical and metallurgical actions occur during switching causing wear, high contact resistance, and welding. Gold or rhodium can be used for currents below 1 A. Silver is used for currents in the 1 to 10 A range for supply voltages above 6 V. Silver cadmium is sometimes used for currents in the range 5 to 25 A when the supply voltage is above 12 V. Mercury wetted contacts are available for currents up to 100 A. The contact life in relays is limited to typically between 100 and 500 K operations.

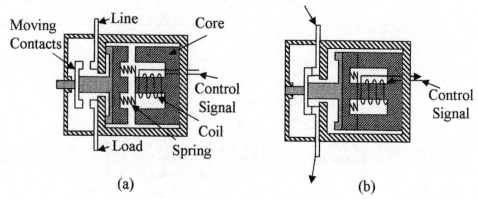

Figure 11.16 A contactor is used for high current and voltage switching, the contactor shown is (a) de-energized and (b) energized.

11.5 Motors

The student needs to be aware of the types of functions the motors perform in industrial applications, but details of motors and control circuits are outside the scope of this text. Motors are used for pumping fluids, compressors, driving conveyer belts, and any form of positioning required in industry. For control applications or positioning, servos or stepper motors are used.

11.5.1 Servo motors

Servo motors can rotate to a given position, be stopped, and reversed. In the case of a servo motor the angular position and speed can be precisely controlled by a servo loop, which uses feedback from the output to the input. Figure 11.17a shows such a system. The position of the output shaft is monitored by a potentiometer which provides an analog feedback voltage to the control electronics (an encoding disc would be used in a digital system), so that the control electronics can use this information to power the output motor and stop it in any desired position or reverse the motor to stop at any desired position.

11.5.2 Stepper motors

Stepper motors rotate at a fixed angle with each input pulse. The rotor is normally a fixed magnet with several poles and a stator with several windings. Eight magnetic poles and a six-section stator are shown in Fig. 11.17b. Stepper motors are available in many different designs with a wide selection of the number of poles and drive requirements, all of which define the stepper motor characteristics and rotation angle for each input phase. Stepper motors can be reversed by changing the sequence of the driving phases. Stepper motors are available with stepping angles of 0.9, 1.8, 3.6, 7.5, 15, and 18 degrees. Since the motor steps a known angle with each input pulse, feedback is not required. However, as only the relative position is known, loss of power will cause loss of position information, so that in a system using stepper motors a position reference is usually required.

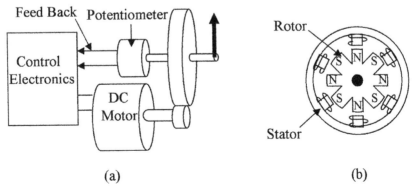

Figure 11.17 Illustrated is (a) a servomotor with a feedback loop and (b) a stepper motor.

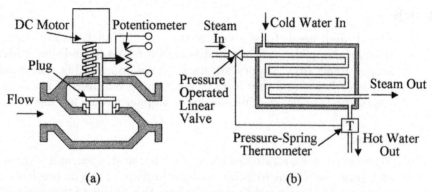

Figure 11.18 (a) A dc electric motor operated valve with position feedback potentiometer and (b) a self-regulating pneumatic temperature controller.

11.5.3 Valve position feedback

In Fig. 11.18a a globe valve operated by an electric motor is shown. The screw driven by the motor can move the plug in the valve up or down. A potentiometer wiper is attached to the valve stem and gives a resistance directly proportional to the amount the valve is open. This resistance value can be fed back to the controlling electronics, so that the position of the valve can be monitored. The system could also be digital, in which case, a digital encoding technique would be used for feedback.

11.5.4 Pneumatic feedback

In Fig. 11.18b pneumatic control is used in a local closed loop system for maintaining water at a set temperature. Cold water and steam are mixed in a heat exchanger; the temperature of the exiting hot water is monitored by a pressure–spring thermometer. The pressure from the thermometer is used to operate and control a linear globe valve in the incoming steam pipe. If the temperature of the hot water increases above a set temperature, the pressure from the thermometer increases and starts to close the valve in the steam line, keeping the hot water at the set temperature. If the flow of hot water increases, the temperature of the water will start to lower and this will reduce the pressure from the thermometer to the valve increasing the steam flow, bringing the temperature back to its set point.

11.6 Application Considerations

11.6.1 Valves

The selection of control valves for a particular application depends on many variables; such as the corrosive nature of the fluid, temperature of operation, pressures involved, high or low flow velocities, volume of flow, and the amount of suspended solids.

Valves are the final element in a control loop and are critical in providing the correct flow for process control. The valve is subject to operation in very harsh conditions and one of the most costly elements in the process control system. The choice and correct installation requires both knowledge and experience. Careful attention must be paid to the system requirements and manufacturers' specifications, only then can a careful valve selection be made (additional information can be obtained from the ISA 75 series of standards).

Some of the factors affecting the choice of valves are as follows:

1. Type of valve for two-way or three-way fail-safe considerations, and so on.
2. Valve size from flow requirements; care must be taken to avoid both oversizing and undersizing.
3. Materials used in the valve construction, considering pressure, size, and corrosion. Materials used in valves range from PVC to brass to steel.
4. *Tightness of shutoff*: Valves are classified by quality of shutoff by leakage at maximum pressure. Valves are classified into six classes depending on leakage from 0.5 percent of rated capacity to 0.15 mL/min. for a 1-in dia. valve.
5. Acceptable pressure drop across the valve.
6. Valve body for linear or rotary motion, i.e., globe, diaphragm versus ball, butterfly, and so forth.

The type of valve or plug depends on the nature of the process reaction. In the case of a fast reaction with small load changes, control is only slightly affected by valve characteristics. When the process is slow with large load changes, valve characteristics are important, i.e., if the load change is linear, a valve with a linear characteristic should be used, in the case of a nonlinear load change, a valve with an equal percentage change may be required. In some applications valves are required to be completely closed when OFF. Other considerations are maintenance, serviceability, fail-safe features, pneumatic, hydraulic, solenoid or motor control, and the need for feedback. The above is a limited review of actuator valves, as previously noted, the manufacturer's data sheets should be consulted when choosing a valve for a particular application.

Position and speed are normally controlled by electric operated servo or stepper motors. In pumping, compressors, conveyer belt, and like applications, three-phase motors are normally used.

11.6.2 Power devices

Power switching devices from contactors to solid state devices will be chosen from considerations of power handling, switching speed, isolation, and cost. Some of the considerations are as follows:

1. For low-speed operation, mechanical relay devices can be used, which will give isolation, relatively low dissipation, and are low cost.

2. Light control and ac motor control can use SCRs and TRIACs. These devices are packaged in a wide range of packages depending on current handling and heat dissipation requirements.

3. For power control, multiphase motor control, and high-speed switching applications BJTs or IGBTs can be used. These devices also come in a variety of low thermal resistance packages.

4. The MOSFET device can be used in medium power applications. The device has the advantage that control circuits can be integrated on to the same die as the power device.

Summary

This chapter discussed the type of valves used to control the manipulated variable, the types of actuators used, and control of power to the actuators.

The main points discussed were as follows:

1. The type of self-regulating gas pressure regulators used in process control, the internal and external loading of the regulators using springs, weights, pressure, and pressure amplifiers.

2. Various methods of automatically controlling liquid levels.

3. A wide variety of control valves are available for flow control. A comparison of their characteristics is given and some options available when choosing a control valve for a specific application are also discussed.

4. Flow control actuators are designed with different control characteristics for different applications such as linear, quick opening valves, and equal percentage valves. The proper characteristics should be chosen for the application.

5. Fail-safe valve configurations are needed to prevent the flow of material during a system failure or loss of power. Valve configurations are shown for valves to fail in the open position or in the closed position.

6. Electronic power control devices are now available for efficient power control with high-speed switching characteristics. The characteristics of the different devices are compared and the control circuits are shown.

7. Magnetic relays and contactors are used for electrical isolation between signal voltage levels and high voltage levels. The devices are used for motor and actuator control.

8. Servo motors and stepper motors are used for position control. It is necessary to feedback to the control system the position of actuator and the like being controlled. Potentiometers for electrical position feedback are shown.

Problems

11.1 What is the prime use of a regulator?

11.2 Where is an actuator used?

11.3 What is an instrument pilot operated pressure regulator?

11.4 What do you understand by fail-safe "open"?

11.5 What are the methods used to load regulators?

11.6 What are the methods used to control actuators?

11.7 Where are electrical contactors used?

11.8 How is the position of a valve communicated back to the controller?

11.9 When are opto-isolators used?

11.10 Where would you use a safety valve?

11.11 Why is a DIAC used in a TRIAC trigger circuit?

11.12 Name the various types of electronic power control devices?

11.13 What are the differences between the SCR and the TRIAC?

11.14 What are the differences between the TRIAC and the IGBT?

11.15 Name the various types of valve families.

11.16 Name the various valve configurations that can be found within the globe valve family.

11.17 A valve has a CV of 88. What is the pressure drop in the valve when 1.8 gal/s of a liquid with a SW of 78 lb/ft^3 is flowing?

11.18 Describe a three-position globe valve.

11.19 In Fig. 11.14, a TRIAC with a 5 V trigger level is used with 12 V zeners. It is required to be able to control the power in the load from full to half power. What is the value of the capacitor C if the potentiometer R_2 is 25 kΩ?

11.20 In Fig. 11.12 the load is 0.5 Ω. If the supply is 120 V ac, what is the maximum power that can be supplied to the load and the power loss in the SCR and diodes? Assume the voltage drop across the SCR and a diode is 1.6 V and 1.5 V, respectively.

Chapter 12

Signal Conditioning

Chapter Objectives

This chapter will help you understand why signal conditioning is required in process control and to familiarize you with signal conditioning methods.

The following are covered in this chapter:

- The conversion of sensor signals into pneumatic or electrical signals
- Signal linearization, methods of setting signal zero level, and span
- Nonlinear analog amplifiers
- Digital linearization
- The difference between sensors, transducers, and converters
- Conditioning for local displays and transmission
- Temperature compensation used in signal conditioning
- Signal conditioning used with Hall effect and magneto resistive element (MRE) devices
- Considerations using capacitive devices
- Resistance temperature detectors (RTD) signal conditioning

Many sensors do not have a linear relationship between the physical variable and the output signal. Output signals need to be corrected for the nonlinearity in their characteristic, or conditioned for transmission to a central controller, or for direct control, so that the necessary valves or actuators can be operated to accurately correct for variations in the measured variable in a process control system.

12.1 Introduction

Sensors are used to convert physical variables into a measurable energy form. This energy form is used to directly or indirectly give a visual indication, as an actuator control signal or as a signal to a controller. Signal conditioning refers to modifications or changes necessary to correct for variations in a sensor's input/output characteristics so that its output bears a linear relationship with the process variable being measured, and the signal is then suitable for use by other elements in the process control loop. Most sensors do not give an output that can directly be used for a visual display or for control. Pressure sensors, for instance, change their shape when pressure is applied giving linear motion, which must then be converted into a dial-type display for direct indication, or an electrical signal for an alpha numeric display. This chapter deals with the conditioning of sensor signals so that they are suitable for use by other linear elements in the system.

Sensors, transducers, and converters were defined in Chap. 1 as follows:

Sensors are devices that sense a variable and give an output (mechanical, electrical, and so on), that is directly related to the amplitude of the variable.

Transducers are systems used to change the output from a sensor into some other energy form so that it can be amplified and transmitted with minimal loss of information.

Converters are used to convert a signal's format without changing the type of energy, i.e., an op-amp that converts a voltage signal into a current signal.

12.2 Conditioning

12.2.1 Characteristics

When choosing a sensor for an application, there is often little choice in the characteristics of the sensor output versus the changes in a process variable. In many cases the relation between the input and the output of a sensor is nonlinear, temperature sensitive, and offset from zero. The situation is aggravated when precise measurements are required and a linear relationship is required between the process variable and the output signal. In analog circuits, linearization is very hard to achieve and requires the use of specialized networks. Figure 12.1a shows the output of a sensor when measuring a variable and the idealized output obtained from a linearization circuit, with adjustment of the gain and bias (zero level) as is required on many types of sensor outputs.

Example 12.1 The output voltage from a sensor varies from 0.35 to 0.7 V as the process variable varies from low to high over its measurement range. However, the sensor output goes to equipment that requires a voltage from 0 to 10 V for the range of the variable. A circuit for changing the output levels is shown in Fig. 12.1b. The negative input to the amplifier is set at 0.35 V to offset the sensor minimum level to give zero out at the low end of the range. The gain of the amplifier is set to 28.6 giving 10 V output with 0.75 V input, i.e., $10/(0.7 - 0.35) = 28.6$. Note the use of impedance matching buffers that would be used in instrumentation.

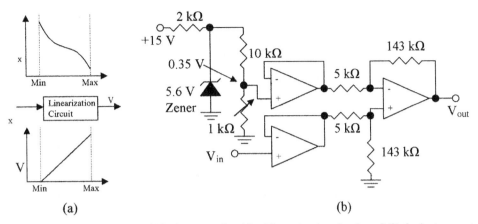

Figure 12.1 (a) The input and ideal output of an ideal linearization circuit and (b) the instrument circuit used for zero and span adjust. This circuit is used in Example 12.1.

Example 12.1 gives a simple example of adjusting for an offset zero. The sensor output could have varied from 0.7 to 0.35 V as the measured variable changed from low to high, and both the offset and span of the sensor could be temperature sensitive. In this case the circuit shown in Fig. 12.2 can be used to invert the signal. The 10 kΩ resistor in the biasing network can be at the same temperature as the sensor and have the same temperature coefficient as the zero offset of the sensor to compensate for zero drift. Span drift or gain can be compensated by a temperature-sensitive resistor in the amplifier feedback. This feedback resistor will also be at the same temperature as the sensor.

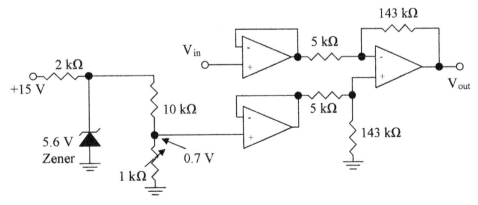

Figure 12.2 Instrument sensor compensation circuit.

12.2.2 Linearization

Example 12.1 shows how to correct for offset. Another problem is nonlinearity in the relation between the measured variable and sensor output. The approach in analog systems and digital systems will be different.

Linearization in analog circuits is difficult unless there is a relatively simple equation to describe the sensor's characteristics. In some applications a much more expensive linear transducer may have to be used due to the inability of analog circuits to linearize the signal conversion. Figure 12.3a shows the circuit of a logarithmic amplifier. Figure 12.3b shows the variations in characteristics with various resistor values that can be obtained for use in signal linearization. When $R_2 = \infty$ and $R_3 = 0$, the amplifier has a logarithmic relation between input and output. When R_3 is larger than zero the gain is higher at the upper end of the scale, as shown. If R_2 is a high value, the effect is to reduce the gain at the lower end of the scale. Multiple feedback paths can be used with nonlinear elements and resistors to approximate the amplifier characteristics to those of the sensor.

Linearization in digital circuits can be performed for nonlinear devices by using equations or memory look-up tables. If the relationship between the values of a measured variable and the output of a sensor can be expressed by an equation, the processor can be programmed on the basis of the equation to linearize the data received from the sensor. An example would be a transducer that outputs a current I related to flow rate v by

$$I = Kv^2 \tag{12.1}$$

where K is a constant.

The current numbers from the sensor are converted into binary, where the relationship still holds. In this case, a linear relationship is required between

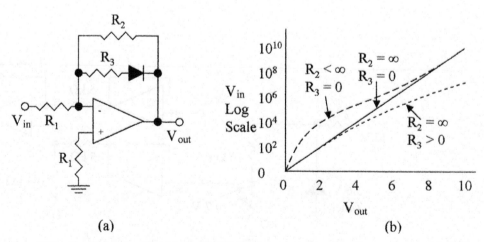

Figure 12.3 Nonlinear amplifier (a) circuit and (b) characteristics of nonlinear circuit with different feedback values.

the current and flow rate. This can be obtained by multiplying the term I by itself, then the resulting number is proportional to v^2, or the generated number and flow now have a linear relationship. Span and offsets may now require further adjustment.

There are many instances in conversion where there is not an easily definable relationship between variable and transducer outputs and it may be difficult or impossible to write a best fit equation that is adequate for linearization of the variable. In this case, look-up tables are used. The tables correlate transducer output and the true value of the variable, and these values are stored in memory so that the processor can retrieve the true value of the variable from the transducer reading by consulting its look-up tables. This method is extensively used with thermocouples.

12.2.3 Temperature correction

Sensors are notoriously temperature sensitive, i.e., their output zero as well as span will change with temperature, and in some cases the change is nonlinear. Variables are also temperature sensitive and require correction. Correction of temperature effects requires a temperature sensitive element to monitor the temperature of the variable and the sensor. The temperature compensation in analog circuits will depend on the characteristics of the sensor used. Because the characteristics of the sensors change from type to type, the correction for each type of sensor will be different. In digital circuits, computers can make the corrections from the sensor and variable characteristics using temperature compensation look-up tables.

Other compensations needed can take the form of filtering to remove unwanted frequencies such as pick up from the 60-Hz line frequency, noise or radio frequency (RF) pickup, dampen out undulations or turbulence to give a steady average reading, correction for time constants, and for impedance matching networks.

12.3 Pneumatic Signal Conditioning

Pneumatic signals as well as electrical signals can be used to control actuators. The bourdon tube, capsule, or bellows convert pressure into mechanical motion which can be used for pneumatic control. Figure 12.4a shows a pneumatic signal conditioner. Air from a 20-psi regulated supply is fed through a constriction to a nozzle and flapper that controls the pressure output. The flapper is mechanically linked to a bellows. When the variable is at its minimum, the linkage opens the flapper, allowing air to be released. The output pressure to the actuator would then be at its minimum, i.e., 3 psi. As the variable increases, the linkage to the flapper causes it to close and the output pressure increases to 15 psi. This gives a linear output pressure range from 3 to 15 psi (20 to 100 kPa) with linear sensor motion and the pressure variations can be used for actuator control. The set zero adjusts the flapper's position and the nozzle can be moved up and down

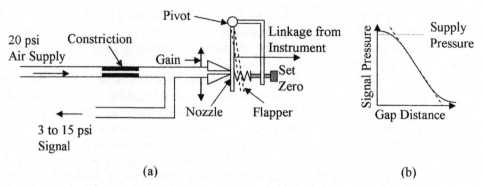

Figure 12.4 Illustrates (*a*) mechanical to pneumatic signal transducer and (*b*) the output pressure versus gap distance.

to give a gain or span control. In some cases the mechanical linkage is reversed so that when the variable is a maximum, the output pressure is 3 psi and 15 psi for the minimum. Figure 12.4*b* shows the relation between the gap distance and the output pressure. The relationship is linear from 3 to 15 psi. Using 3 psi as a minimum gives an additional advantage in that 0 psi indicates a fault condition. Newer systems will use electrical signals in preference to pneumatic signals, as no pressure line, regulator, or compressor is required. Pneumatic control is not compatible with microcontrollers.

12.4 Visual Display Conditioning

The method of signal conditioning can vary, depending on the destination of the signal. For instance, a local signal for a visual display will not require the accuracy of a signal used for process control. Visual displays are not normally temperature compensated or linearized. They often use mechanical linkages which are subject to wear over time giving a final accuracy between 5 and 10 percent of the reading, i.e., there is little or no conditioning. However, with most of the nonlinear sensors, the scale of the indicator will be nonlinear to give a more accurate indication. These displays are primarily used to give an indication that the system is working within reasonable limits or is within broadly set limits, i.e., tire pressures, air conditioning systems, and the like.

12.4.1 Direct reading sensors

A few sensors have outputs that are suitable for direct reading at the point of measurement, but the outputs cannot be used for control or transmission. Such devices are sight glasses for level indication, liquid in glass for temperature, a rotameter for flow, hydrometer for density or specific gravity (SG), and possibly a liquid filled U-tube manometer for differential or gauge pressure measurements.

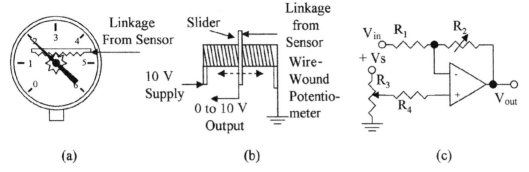

Figure 12.5 Mechanical linkage for (a) direct reading indicator, (b) wire-wound potentiometer, and (c) a simple circuit for use with the potentiometer.

Visual indicators should be clear for ease of reading and the scale well defined. Rotameters need to be selected for flow rates, fluid density, and their output value should be corrected for temperature variations from look-up tables. Care needs to be taken to ensure that the thermometer bulbs are correctly placed in the fluid for temperature measurement and do not touch the container walls, as this can affect the temperature reading. When measuring liquid levels and liquid and gas pressures, the instrument should have conditioning baffles to minimize pressure and level fluctuations which can introduce uncertainties into the readings.

The Bourdon tube, capsule, or bellows convert pressure into mechanical motion which is well suited for conversion to direct visual indication; the Bourdon tube for instance is normally an integral part of the indicator. Figure 12.5a shows a mechanical linkage from a sensor to a direct reading indicator as is normally used for pressure sensing. The Bourdon tube is normally located behind the dial. As the pressure changes the Bourdon tube changes its radius and moves the toothed slider to operate the pointer. The pointer moves over a scale which is graduated in pounds per square inch and so on. These devices are cost effective and are in wide scale use, but are not temperature compensated and the cheaper instruments do not have a zero or span adjustment. More expensive devices may have screw adjustments and a limited temperature range.

12.5 Electrical Signal Conditioning

The accuracy of the sensor signal is not only dependent on the sensor characteristics but mainly on the applied conditioning. Many processes require variables to be measured to an accuracy of more than 1 percent over the full range, which means not only very accurate sensing, but also temperature compensation, linearization, zero set, and span adjustment. Temperature compensation is achieved in many sensors by using them in bridge circuits but further compensation may be needed to correct for changes in the variable due to temperature. Such things as op-amp offset and amplification are affected by supply voltages so that these will have to be regulated and care must be taken with grounding of the system to minimize noise and zero offset. Careful selection is needed in the choice of components. Quality close tolerance components and the use of impedance-matching devices are required to prevent the introduction of errors in conditioning networks.

12.5.1 Linear sensors

Figure 12.5b shows a mechanical linkage from a sensor to the wiper of a potentiometer. In this case the variable is converted into an electrical voltage, giving a voltage output from 0 to 10 V. The output voltage can be fed to a voltmeter, converted to a current with an amplifier, or digitized to operate a remote sensing indicator, an actuator, or a signal to a controller. Figure 12.5c shows a circuit that could be used for conditioning with set zero and gain control potentiometers. The set zero can be adjusted by R_3 to give zero output with minimum input and the span adjusted by R_2 to give the required gain. The supply voltage to the amplifier and +Vs to R_3 will need to be regulated voltages. However impedance matching devices should be used in instrumentation.

Figure 12.6a gives an alternative method of signal conditioning the linear motion output from a bellows into an electrical signal using a linear variable differential transformer (LVDT). The bellows converts the differential pressure between P_1 and P_2 into linear motion, which changes the position of the core in the LVDT.

Figure 12.6b shows a circuit that can be used to condition the electrical signal output from an LVDT. As the output from the transformer is ac, diodes are used to rectify the signal. The signal is then smoothed using a resistor-capacitor (RC) filter, and the two dc levels are fed to an op-amp for comparison. The set zero and span adjustments are not shown.

12.5.2 Float sensors

A float is often used for level measurements. The level can be converted into angular or linear motion, but gives a somewhat nonlinear output, as many of

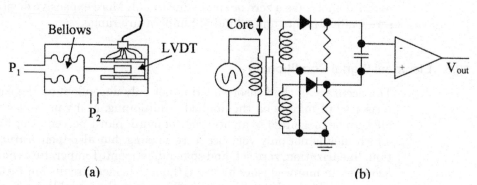

Figure 12.6 (a) Differential pressure bellows converting pressure into an electrical signal using an LVDT and (b) a signal conditioning circuit for the LVDTs.

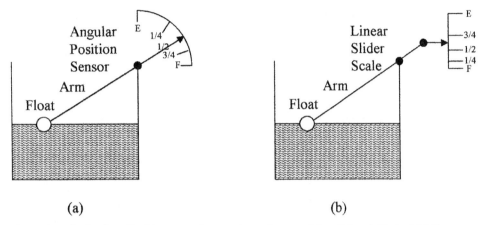

Figure 12.7 Scales for a float type angular arm type of sensor (a) radial and (b) straight line.

us have found with our automotive gas gauge. Figure 12.7a shows the relation between fluid level and rotational scale. The scale is cramped at the full end, so that the output from a rotational type potentiometer will give small voltage changes when the container is full, compared to large voltage changes when the container is approaching empty. Figure 12.7b shows the relation between fluid level and a linear scale, which is also cramped at the full end, so that in either case the scales are similar.

Example 12.2 Figure 12.8a shows a float connected to a 10 kΩ potentiometer with a 10 V supply. Calculate the output voltage for liquid levels from empty to full, plot the relationship, and estimate the accuracy from the best fit straight line. If when the container was half-full as indicated by the potentiometer shown in Fig. 12.8b, the output voltage was adjusted to be half the supply voltage using a resistor from V_{out}, what would be the resistor value and would it be connected to 0 or 10 V? What would be the accuracy from the best fit straight line with the resistor?

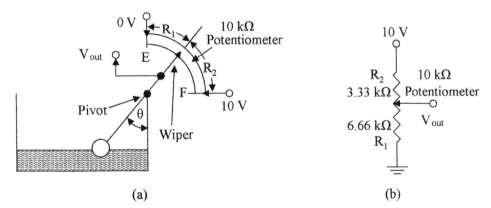

Figure 12.8 For use with Example 12.2a shows float connected to a 10-kΩ potentiometer and (b) 10 kΩ potentiometer readings at liquid half-full.

TABLE 12.1 Resistance Values, Angle, and V_{out} for Various Liquid Levels with Float Sensor

Fluid level	E	1/8	1/4	1/2	3/4	F
$\cos^{-1}\theta$	0	7/8	3/4	1/2	1/4	1
Θ degrees	0	28.9	41.4	60	75.5	90
R_1 kΩ	0	3.2	4.6	6.66	8.3	10
R_2 kΩ	10	6.8	5.4	3.33	1.7	0
V_{out}	0	3.2	4.6	6.7	8.3	10

From Fig. 12.8a it is possible to calculate the values of θ, R_1, R_2, and V_{out} for various liquid levels. These are shown in Table 12.1.

Figure 12.9a shows a plot of the output voltage versus the volume of liquid in the tank from Table 12.1 for the uncompensated curve. The best-fit straight line (dashed) gives an error of about ±15 percent of full scale reading (FSD). To correct the mid-point of the scale for 5 V output when the tank is half full (Fig. 12.8b), the circuit would require a 6.66-kΩ resistor from the wiper to ground as shown in Fig. 12.9b (the resistor should be made with the same material as the potentiometer). Table 12.2 gives the liquid levels, resistance values with the potentiometer in parallel with the 6.66 kΩ resistor, and the output voltage for the circuit in Fig. 12.9b. The new values are also plotted in Fig. 12.9a as the compensated curve. The best-fit straight line (dashed) gives an error of less than ±5 percent showing the improved linearity with simple conditioning. The output voltage from the float sensor can also be compensated by the control processor before being fed to other elements in the system.

An alternative to the float attached to an arm is a float with a counter balance as shown in Fig. 12.10a. This arrangement will give a linear scale with liquid level or if a rotary potentiometer is attached to the pointer pivot, the output voltage from the potentiometer will be linear with liquid level as shown in Fig. 12.10b.

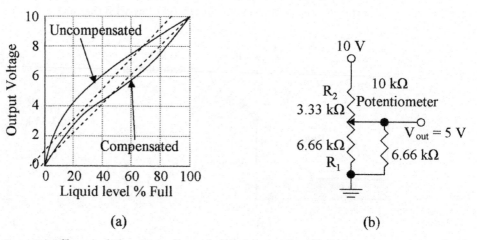

Figure 12.9 Characteristic output voltage (a) plotted against liquid level for an uncompensated and compensated float sensor and (b) the circuit with a compensated potentiometer.

TABLE 12.2 Compensated Resistance Values and V_{out} for Liquid Levels

Fluid level	E	1/8	1/4	1/2	3/4	F
R_1 in parallel with 6.66 kΩ	0	2.14	2.69	3.3	3.65	3.94
R_2	10	6.8	5.4	3.3	1.7	0
V_{out}	0	2.4	3.3	5	7	10

12.5.3 Strain gauge sensors

Diaphragms use strain gauge or capacitive sensing, the movement being too small to control a pneumatic flapper, slider, or potentiometer. The strain gauge elements are resistors made from copper or nickel particles glued onto a non-conducting substrate; semiconductor strain gauges are also available that use the piezoresistive effect.

A strain gauge normally consists of two strain elements mounted at right angles to each other and in close proximity, so that they are both at the same temperature. See Fig. 12.11a. The gauge is mounted on the diaphragm with one gauge in line with the direction of maximum strain for strain measurement and the other perpendicular to the line of strain, so that it will not sense the strain, and is used to provide temperature compensation and signal conditioning for the strain gauge element when used in a bridge circuit.

Figure 12.11b shows a circuit using the strain gauge. The strain gauge elements are mounted in two arms of the bridge and two resistors, R_1 and R_2, form the other two arms, R_3 and R_5 are the conditioning for the zero offset and span, respectively. The output signal from the bridge is amplified and impedance matched, as shown. The strain gauge elements are in opposing arms of the bridge, so that any change in the resistance of the elements due to temperature changes will not affect the balance of the bridge, giving temperature compensation.

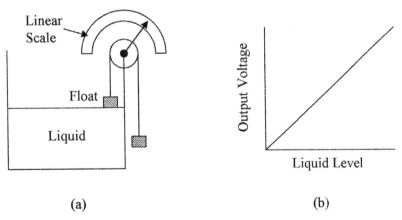

Figure 12.10 (a) Float-type sensor with a linear radial scale and (b) output voltage versus liquid level when the scale is replaced by a potentiometer.

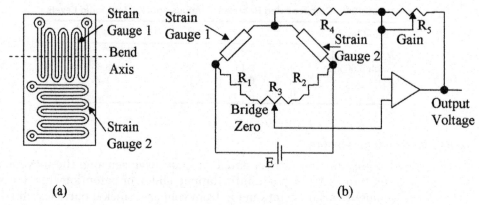

Figure 12.11 (a) Configuration for strain gauge elements and (b) Resistive bridge for signal conditioning of a strain gauge.

More gain and impedance matching stages than shown may be required or an A/D converter will be required to make the signal suitable for transmission. Additional linearization may also be needed; this information can be obtained from the manufacturers' device specifications.

12.5.4 Capacitive sensors

Capacitive sensing devices can use single-ended sensing or differential sensing. With single-ended sensing, capacitance is measured between the diaphragm and a single capacitor plate in close proximity to the diaphragm. Differential sensing can be used when there are capacitor plates on either side of, and in close proximity to, the diaphragm (see Fig. 12.12a). In differential sensing the two capacitors can be used to form two arms of an ac bridge or switch capacitor techniques

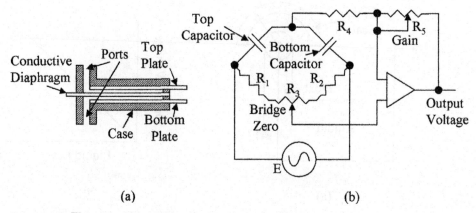

Figure 12.12 Illustrates (a) capacitive diaphragm pressure sensor and (b) an ac bridge for use with a capacitive sensor.

can be used. For single ended sensing a fixed reference capacitor can be used. Capacitive sensing can use ac analog or digital techniques. Microminiature pressure sensors can use piezoresistive strain gauge sensing or capacitive sensing techniques. Being a semiconductor-based technology, the sensor signal is conditioned on the die, i.e., amplified, impedance matched, linearized, and temperature compensated.

Figure 12.12b shows an ac bridge with offset and span conditioning that can be used with capacitive sensing. Initially, the bridge is balanced for zero offset with potentiometer R_3, the output from the bridge is amplified and buffered. The output amplitude can be adjusted by the gain control, R_5. The signal will need to be converted to a dc signal and further amplified for transmission.

12.5.5 Resistance sensors

Resistive temperature detectors (RTD) measure the change in the electrical resistance of a wire-wound resistor with temperature, typically, a platinum resistance element is used with a resistance of about 100 Ω. The resistance change can be measured in a bridge circuit, but normally the resistor is driven from a constant current source and the voltage developed across the resistor measured. Care must be taken with these devices to ensure that the current flowing through the devices is low to minimize the temperature changes occurring due to the internal heating of the resistor. Pulse techniques can be used to prevent internal heating. In this case the current is turned on for a few milliseconds, the voltage measured and then turned off for, say, a second. Figure 12.13a shows the simplest connection to the RTD with just two leads, the meter being connected to the current supply leads. The resistance of long leads between the detector and the resistor contribute to measurement error, as the meter is measuring the voltage drop across the current lead resistance and junctions as well as the RTD.

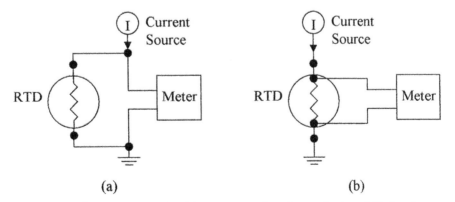

Figure 12.13 RTD connections using (a) common supply and meter leads and (b) directly connected meter.

Figure 12.13b shows a 4-wire connection to an RTD. The meter connects directly to the RTD so that only the voltage drop across the RTD is measured; no error is introduced due to the resistance of the current supply contacts or lead resistance to the RTD. Platinum is the material of choice for an RTD. A linearity of 3.6 percent can be obtained from 0 to 850°C without signal conditioning. The RTD has a constant current flowing through it giving an offset zero, so that zero level correction and span conditioning are required. Direct connection of the resistive element to the controller will be further discussed in Chap. 13.

12.5.6 Magnetic sensors

Many flow measurements are sensed as differential pressures with the indicator scale graduated in cubic feet per minute, gallons per minute, liters per second, and so forth. Rotating devices, such as the turbine, are used for accurate flow measurements. The devices are simple, do not require conversion to pressure or other medium, have low drag, can be constructed of inert materials that are resistant to corrosion, do not require regular recalibration, and are low maintenance. The pick off is a magnetic sensor such as a Hall effect or an MRE device. The Hall device gives an electrical impulse every time a blade passes under the sensor, whereas the resistance of the MRE device changes in a changing magnetic field. Figure 12.14a shows the circuit used to shape the signal from an MRE into a digital signal. The MRE sensor contains four elements to form a bridge circuit as shown. The Hall or MRE device does not normally require temperature compensation as they are being used as switches in digital configurations. To measure flow rates a window is opened for a known time period. The number of impulses from these devices are counted from which the rate of flow can be calculated. These devices can also be used to measure the total volume in gallons or liters; in this case the number of pulses from the sensors can be counted

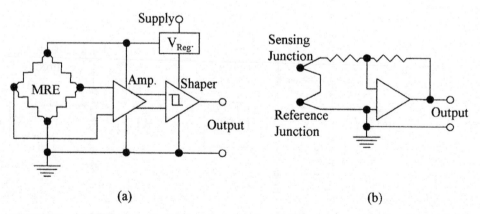

Figure 12.14 (a) MRE magnetic field sensing circuit and (b) a thermocouple signal amplifying circuit.

and divided by the number of pulses per unit volume to give the total volume that has flowed. Some conditioning may be required for the density changes in the liquid with temperature and for high and low flow rates. The conditioning will depend on the requirements of the application and manufacturers' specifications.

12.5.7 Thermocouple sensors

Thermocouples are connected as shown in Fig. 12.14b. The sensing junction and the reference junction are connected in series. When the junctions are at the same temperature the voltage output from the junctions is zero, and the output from the amplifier is zero. When the junctions are at different temperatures, there is a differential voltage at the input to the amplifier that is amplified and converted to a temperature reading. To make this an absolute reading the temperature of the reference junction is required. This can be achieved by placing the junction in a constant temperature enclosure, or the temperature of the reference junction can be measured and a correction applied to the output reading of the thermocouple's sensing amplifier. The amount of conditioning required by a thermocouple will depend on its temperature measuring range (see Table 8.6). Its accuracy is 1 to 2 percent over a limited temperature range but needs conditioning (linearizing) if used over its full operating range. The set zero conditioning is a part of the reference temperature correction. Thermocouple voltages can also be sensed directly by the controller using an internal amplifier and then conditioned internally; this is discussed in Chap. 13.

12.5.8 Other sensors

Piezoelectric sensors are used for sensing force that can be produced by pressure, weight, or acceleration. These devices have high sensitivity but are poor at sensing low-level forces due to offset and drift caused by temperature variations. Piezoelectric devices are normally packaged with a buffer amplifier and conditioning as shown in Fig. 5.12b but extra gain stages may be needed before transmission. Any extra conditioning necessary can be obtained from the manufacturers' specifications.

Angular and distance measuring devices are digital in nature so that any conditioning required is done at the controller; only the correct format is needed for transmission.

Bimetallic sensors are temperature-sensing devices that can be used with a mechanical linkage to operate a display directly such as in an oven thermometer but are not accurate enough for linear control applications. Other bimetallic devices operate switches (mercury in glass or mechanical contacts) as simple / ON/OFF devices; such are used in thermostat applications normally without conditioning.

Taguchi, chemical, smoke detectors, and similar types of sensors all require conditioning; the type and extent of the conditioning required depends on the application and manufacturers' specifications and will not be discussed here.

12.6 A-D Conversion

Many analog signals are converted to digital signals for transmission. In many cases the output from a sensor can be converted directly into a digital signal, such as with capacitive sensors. The value of capacitors can be accurately sensed using digital techniques, eliminating the need for analog amplification, but conditioning of the sensor is still required. Resistive-type devices can also be sensed directly using digital techniques, but again temperature compensation of the sensor is required. However, in the digital domain, all the conditioning can be performed by the processor in the controller, using software or look-up tables.

Summary

This chapter introduced signal conditioning and some of the various methods used to linearize signals and the correction needed for zero offset.

The salient points covered were as follows:

1. The methods used to convert the displacement sensor signal into a pneumatic or electrical signals. The transducer used can be a potentiometer, LVDT, or capacitive devices.
2. Linearization of signals, changing operating levels and gain control using non-linear analog amplifiers or equations and look-up tables when using digital methods.
3. Pneumatic signal conditioning and system failure detection.
4. Conditioning for direct reading visual displays.
5. Conversion of sensor signals to electrical signal for conditioning.
6. Techniques used in the temperature compensation of strain gauges and other types of devices.
7. Capacitive sensor measurements using ac bridge circuits.
8. The methods used to reduce errors in the measurement of RTD signals.
9. Magnetic sensors are digital in nature and do not require the same type of conditioning as analog devices but can be nonlinear and conditioning of the sensor and material being measured may be required.

Problems

12.1 Name two magnetic field sensors.

12.2 What is the difference between a sensor, a transducer, and a converter?

12.3 Why is it necessary to use signal conditioning on sensor signals?

12.4 How do pneumatic transducers control pressure?

12.5 The output voltage from a sensor varies from a minimum of 0.21 V to a maximum of 0.56 V. Draw a circuit to condition the signal so that the output voltage goes from 0 to 10 V. Assume a reference voltage of 10 V, the resistor from the reference is 10 kΩ and the input resistor to the amplifier is 5 kΩ.

12.6 What methods are used to convert mechanical sensor movement into electrical or display signals?

12.7 What are the methods of providing the reference for a thermocouple?

12.8 Why are strain gauges normally mounted in pairs at right angles to each other?

12.9 What types of transducers are used with diaphragm pressure-type sensors?

12.10 A float sensor using a 27 kΩ angular position potentiometer is used to measure liquid level in a tank, see Fig. 12.5. What is the optimum value of resistance between the wiper and ground to compensate for the nonlinearities in the system?

12.11 What types of analog circuits are used for linearization?

12.12 What is the effect of a resistor in series with a nonlinear element?

12.13 What is the effect of a resistor in parallel with a nonlinear element?

12.14 How is linearization performed in digital circuits?

12.15 How are temperature corrections made to temperature sensitive sensors?

12.16 What are the pressure ranges used in pneumatic signal transmissions? Why is zero not used?

12.17 How can you use a float level sensor to obtain a linear relation between level and electrical output?

12.18 How is the output from a capacitive type of sensor measured?

12.19 Why should the meter contacts to an RTD be as close to the measuring element as possible?

12.20 Which types of sensors are suitable for direct line of sight reading but are hard to use with a transducer?

Chapter 13

Signal Transmission

Chapter Objectives

This chapter will help you understand the prime modes of signal transmission and familiarize you with the various methods of signal transmission and where they are used.

This chapter discusses the following:

- Pneumatic signal transmission
- Types of analog electrical signal transmission
- Electrical to pneumatic signal converters
- Thermocouple and resistive type devices and temperature signal transmission
- The operation of the signal processor in signal transmission
- Smart sensors and fieldbus
- Programmable logic controllers (PLCs) and ladder diagrams
- Telemetry signal transmission
- Conversion of digital signals into analog signals for actuator control

Measurement of variables are made by sensors, conditioned by transducers, and then transferred to another location using a transmitter. In the case of process control, the accuracy of transmission of the value of the variable is very important; any errors introduced during transmission will be acted upon by the controller and will degrade the accuracy of the signal. There are several methods of transmitting data. The chosen solution will depend on the sensor, application of the signal, the distance the signal needs to be sent, the accuracy requirements of the system, and cost. Unfortunately, the accuracy of the system can be degraded by poor transmission.

13.1 Introduction

The various methods of signal transmission are discussed in this chapter. Control signals can be transmitted pneumatically or electrically. Due to the needs of an air supply for pneumatic transmission, inflexible pluming, cost, slow reaction time, limited range of transmission, reliability, accuracy, and the requirements of control systems, electrical transmission is now the preferred method. Electrical signals can be transmitted in the form of voltages, currents, digitally, optically, or via wireless. Unfortunately, the terms transducer, converter, and transmitter are often confused and used interchangeably.

Transmitters are devices that accept low-level electrical signals and format them, so that they can be transmitted to a distant receiver. The transmitter is required to be able to transmit a signal with sufficient amplitude and power so that it can be reproduced at a distant receiver as a true representation of the input to the transmitter, without any loss of accuracy or information.

Offset refers to the low end of the operating range of a signal. When performing an offset adjustment, the output from the transducer is being set to give the minimum output (usually zero) when the input signal value is a minimum.

Span refers to the range of the signal, i.e., from zero to full-scale deflection. The span setting (or system gain) adjusts the upper limit of the transducer with maximum signal input. There is normally some interaction between offset and span; the offset should be adjusted first and then the span.

13.2 Pneumatic Transmission

Pneumatic signals were used for signal transmission and are still in use in older facilities or in applications where electrical signals or sparks could ignite combustible materials. Pneumatic transmission of signals over long distances require an excessively long settling time for today's processing needs and when compared to electrical signal transmissions. Pneumatic signal lines are also inflexible, bulky, and costly compared to electrical signal lines and are not microprocessor compatible. Hence, they will not be used in new designs, except possibly, in special circumstances as mentioned. Pneumatic transmission pressures were standardized into two ranges, i.e., 3 to 15 psi (20 to 100 kPa) and 6 to 30 psi (40 to 200 kPa); the 3 to 15 psi is now the preferred range. Zero is not used for the minimum of the ranges as low pressures do not transmit well and the zero level can then be used to detect system failure.

13.3 Analog Transmission

13.3.1 Noise considerations

Analog voltage or current signals are hard wired between the transmitter and the receiver. Compared to digital signals, these signals can be relatively slow to settle due to the time constant of the lead capacitance, inductance, and resistance, but are still very fast in terms of the speed of mechanical systems. Analog signals can loose accuracy if signal lines are long with high resistance; can be susceptible to ground offset, ground loops, noise and radio frequency pick-up. Figure 13.1a shows the controller supplying dc power to the transmitter and the signal path from the transmitter to the controller. The dc power for the sensors can be obtained from the controller to save the cost of deriving the power at the sensor as shown in Fig. 13.1b. However, the current flowing in the ground line (shown in Fig. 13.1a) from the supply will be much larger than the signal current and will produce a voltage drop across the resistance of the ground lead, elevating the ground level of the transmitter which will give a signal offset error at the controller. The second problem with this type of hard wiring is that it is susceptible to radio frequency (RF) and electromagnetic induction (EMI) noise pick-up, i.e., the induced noise from RF transmitters and motors will produce error signals.

To reduce these problems the setup shown in Fig. 13.1b can be used. This setup shows that the dc supply to the transmitter is generated from the ac line voltage via an isolation transformer and voltage regulator at the transmitter. The ground connection is used only for the signal return path. The signal and ground return leads are a screened twisted pair, i.e., the signal leads are screened by a grounded sheath. The RF and EMI pickup are reduced by the screen and the induced noise in both lines is greatly reduced. Because variations in the supply voltages can produce changes in the offset voltage and the gain of the sensor/transmitter, the supply voltage must be regulated.

An improved method of minimizing RF and EMI pickup is shown in Fig. 13.2. In this case, the transmitter sends a differential signal using a screened twisted pair. The reduced pickup will affect both signals by the same amount and will cancel in the differential receiver in the controller. Differential signals are not normally affected by ground offsets.

A differential output voltage signal can be generated using the circuit shown in Fig. 13.3. The output stages have unity gain to give low output impedance and equal and opposite phase signals. Op-amps are also commercially available with differential outputs which can be used to drive buffer output stages.

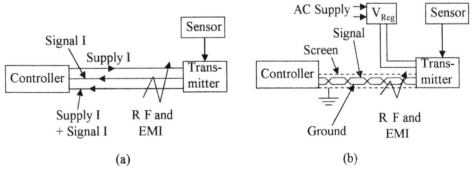

Figure 13.1 Supply and signal connections are shown between controller and transmitter using (a) straight leads and (b) a twisted pair.

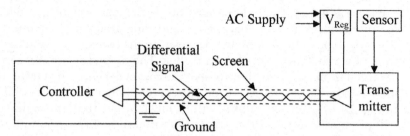

Figure 13.2 Screened differential signal connection between the controller and the transmitter.

13.3.2 Voltage signals

Voltage signals are normally standardized in the voltage ranges 0 to 5 V, 0 to 10 V, or 0 to 12 V, with 0 to 5 V being the most common. The requirements of the transmitter are a low output impedance to enable the amplifier to drive a wide variety of loads without a change in the output voltage, low temperature drift, low offset drift, and low noise. Figure 13.4a shows a transmitter with a voltage output signal. Its low output impedance enables the driver to charge up the line capacitance, achieving a quick settling time. However, the input voltage to the controller V_{in} can be less than the output voltage V_{out} from the transmitter due to resistance losses in the cables if the receiver is drawing any current, i.e.,

$$V_{in} = \frac{V_{out} \times \text{Internal } R}{\text{Internal } R + 2 \times \text{Wire } R} \tag{13.1}$$

The internal R of the controller must be very high compared to the resistance of the wire and connections, to minimize signal loss (which is normally the case).

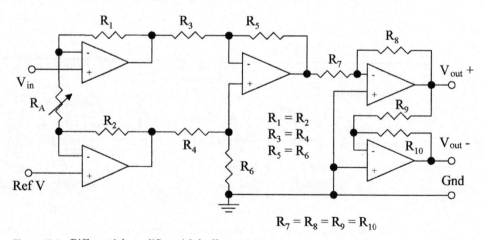

Figure 13.3 Differential amplifier with buffer outputs.

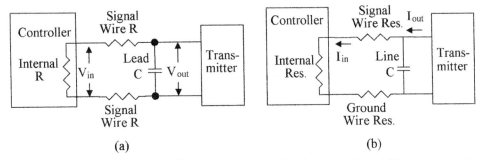

Figure 13.4 Effect of resistance and lead capacitance on (a) voltage signals and (b) current signals.

A differential signal as shown in Fig. 13.4 will eliminate ground noise and offset problems.

13.3.3 Current signals

Current signals are standardized into two ranges; these are 4 to 20 mA and 10 to 50 mA, where 0 mA is a fault condition. The latter range was the preferred standard but has now been dropped, and the 4 to 20 mA range is the accepted standard. The requirements of the transmitter are high output impedance, so that the output current does not vary with load, low temperature, offset drift, and low noise. Figure 13.4b shows a transmitter with a current output. The main disadvantage of the current signal is its longer settling time due to the high-output impedance of the driver which limits the available current to charge up the line capacitance. After the line capacitance is charged, the signal current at the controller is the same as the signal current from the transmitter and is not affected by normal changes in lead resistance. The internal resistance of the controller is low for current signals, i.e., a few hundred ohms. Again a differential signal connection eliminates noise and ground problems.

13.3.4 Signal conversion

Signal conversion is required between low-level signals and high-energy control signals for actuator and motor control. Control signals can be either digital, analog voltage or analog current, or pneumatic. It is sometimes necessary to convert electrical signals to pneumatic signals for actuator control. Pneumatics is still used in applications where the cost of converting to electrical control would be prohibitive, electromagnetic (e/m) radiation could cause problems, or in a hazardous environment where sparks from electrical devices could cause volatile material to ignite.

A linear pneumatic amplifier or booster can be used to increase the pressure from a low-level pressure signal to a high pressure signal to operate an actuator. Figure 13.5a shows a pressure amplifier. Gas from a high-pressure supply is controlled by a conical plug which is controlled by a diaphragm whose position is set

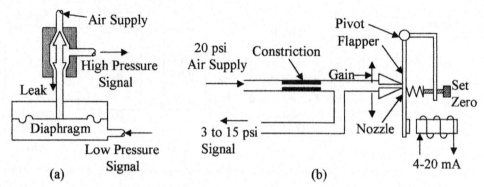

Figure 13.5 Signal conversion (*a*) pressure amplifier and (*b*) current to pressure transducer.

by a low-pressure signal. The gain of the system is set by the area of the diaphragm divided by the area of the base of the conical plug. The output pressure is inverted but linear with respect to the input pressure; the device shown is one of many different types. Pneumatic feedback can be used to improve the characteristics of the amplifier.

One of the many designs of a current-to-pressure converter is shown in Fig. 13.5*b*. The spring tends to hold the flapper closed, giving a high-pressure output (15 psi). When current is passed through the coil the flapper moves towards the coil opening the air gap at the nozzle reducing the output air pressure. The output air pressure is set to the maximum of 15 psi by the set zero adjustment when the current through the coil is 3 mA. The system gain and span is set by moving the nozzle along the flapper. The output pressure is inverted with respect to the amplitude of the current in the setup as shown, but could be set up to be noninverting. There is a linear relationship between current and pressure.

13.3.5 Thermocouples

Thermocouples have several advantages over other methods of measuring temperature, in that they are very small in size, have a low time response (10/20 ms compared to several seconds for some elements), are reliable, have good accuracy, a wide operating temperature range, and they can convert temperature directly into electrical units. The disadvantages are the need for a reference and the low signal amplitude. Thermocouple signals can be amplified with a cold junction reference close to the amplifier and the signal transmitted in an analog or digital format to a controller, or the thermocouple can be directly connected to the controller for amplification and cold junction correction. This method is sometimes used to eliminate the cost of remote amplifiers and power supplies. Controller peripheral modules are available for amplification of several thermocouple inputs with cold junction correction; Fig. 13.6*a* shows a differential connection between the amplifier and the thermocouple as a twisted pair of wires that is screened to minimize noise and the like. Figure 13.6*b* shows

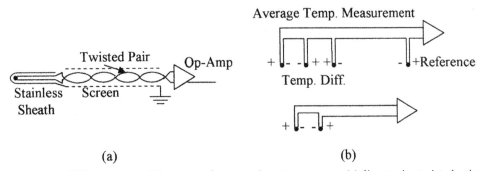

Figure 13.6 Different types of thermocouple connections to an op-amp (a) direct using twisted pair to a reference and amplifier and (b) for average temperature measurement and differential temperature measurement.

other configurations that can be used to connect thermocouples for temperature averaging and differential temperature measurements.

13.3.6 Resistance temperature devices

Resistance temperature devices (RTD) can be connected directly to the controller peripheral amplifiers using a two-, three-, or four-wire lead configuration; these are shown in Fig. 13.7. The RTD is driven from a constant current source I and the voltage drop across the RTD measured. The two-wire connection (a) is the simplest and cheapest, the three-wire connection (b) is a compromise between cost and accuracy, and the four-wire connection (c) is the most expensive but most accurate. The wires in all cases will be in screened cables. In the case of the two-wire connection the voltage drop is measured across the lead wires as well as the RTD; the resistance in the two-lead wires can be significant, giving a relatively high degree of error.

In the case of the three-wire connection, a direct return lead from the RTD to the voltmeter is added, as shown. The voltage drop δV between the ground connection and the lower RTD connection as well as the voltage drop V between

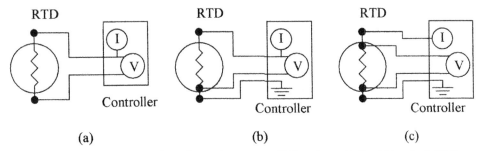

Figure 13.7 Alternative connection schemes between an RTD and a controller (a) two lead, (b) three lead, and (c) four lead.

the current source and the lower RTD connection can be measured. If the resistance in each supply lead to the RTD is assumed to be the same, the voltage across the RTD is $V - \delta V$ correcting for the error caused by the common lead wire. In most cases each lead wire will have about the same resistance, so this method is accurate enough for most applications. With the four-wire connection the voltmeter is connected directly to the RTD as shown in Fig. 13.7c and because no current flows in the leads to the voltmeter there is no voltage drop in the measuring leads and an accurate RTD voltage reading is obtained.

13.4 Digital Transmission

13.4.1 Transmission standards

Digital signals can be transmitted via a hardwired parallel or serial bus, radio transmission or fiber optics, without loss of integrity. Digital data can be sent faster than analog data due to higher speed transmission. Another advantage is that digital transmitters and receivers require much less power than analog transmission devices.

Communication standards for digital transmission between computers and peripheral equipment are defined by the Institute of Electrical and Electronic Engineers (IEEE). The standards are the IEEE-488 or RS-232. However, several other standards have been developed and are now in use. The IEEE-488 standard specifies that a digital "1" level will be represented by a voltage of 2 V or greater and a digital "0" level shall be specified by 0.8 V or less as well as the signal format to be used. The RS-232 standard specifies that a digital "1" level shall be represented by a voltage of between +3 V and +25 V and a digital "0" level shall be specified by a voltage of between −3 V and −25 V as well as the signal format to be used. Fiber optics are now also being extensively used to give very high speed transmission over long distances and are not affected by electromagnetic or RF pickup. Figure 13.8 shows a two way fiber optic cable set up with light emitting diode (LED) drivers and photodiode receivers.

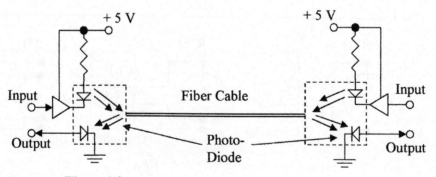

Figure 13.8 Fiber-optic bus.

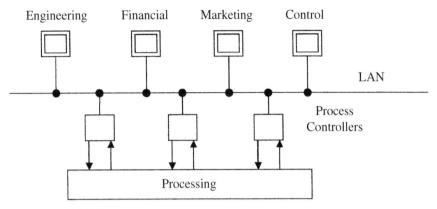

Figure 13.9 A LAN network.

Digital signals can be transmitted without loss of accuracy and can contain error correction codes for limited automatic error correction or to automatically request data retransmission. These networks are known as local area networks (LAN) when used in a limited area such as a plant or wide area networks (WAN) when used as a global system. A typical LAN network is shown in Fig. 13.9. Engineering, finance, and marketing can communicate with the process controllers to monitor plant operations for cost figures and product delivery details over the LAN, directly from the process control system.

Computer based process control systems are flexible systems with a central processor and the ability to add interface units on a limited basis. The interface units can be receivers for reception of analog and/or digital information from the monitoring sensors or transmitters for sending control information to control actuators. A typical receiver unit will contain 8 analog amplifiers with analog to digital convertors (ADCs) giving the unit the ability to interface with 8 analog transmitting devices and change the data into a digital format to interface with the processor. Other interface units contain thermocouple amplifiers or bridges for use with resistive sensors. A data transmitter unit will have the capability of controlling 8 actuators and will contain 8 digital-to-analog convertor (DAC) to change the digital data to an analog format for each actuator being controlled. This set up is shown in Fig. 13.10. Each input or output requires its own interconnect cable or bus resulting in a mass of wiring which requires careful routing and identification marking.

13.4.2 Smart sensors

Smart sensor is a name given to the integration of the sensor with an ADC, processor, and DAC for actuator control and the like; such a setup for furnace

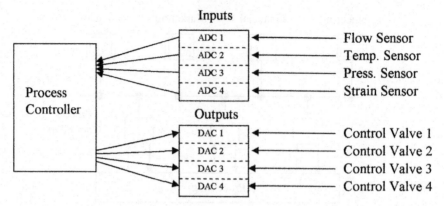

Figure 13.10 Process system with individual inputs and outputs for each variable.

temperature control is shown in Fig. 13.11. The electronics in the smart sensor contains all the circuits necessary to interface to the sensor, amplify the signal, apply proportional, integral, and derivative (PID) control (see Chap. 14), sense temperature to correct for temperature variations in the process if required, correct for sensor nonlinearity, the ADC to convert the signal into a digital format for the internal processor, and the DAC to convert the signal back into an analog format for actuator control. The processor has a serial digital bus interface for interfacing via the fieldbus to a central computer. This enables the processor in the smart sensor to receive update information on set points, gain, operating mode, and so on, and to send status information to the central computer.

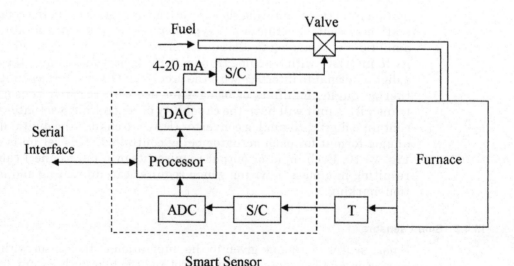

Figure 13.11 Smart sensor block diagram.

13.4.3 Foundation Fieldbus and Profibus

Foundation Fieldbus (FF) and Profibus are the two most universal serial data bus formats that have been developed for interfacing between a central processor and smart sensing devices in a process control system. The Foundation Fieldbus is primarily used in the United States and the Profibus format is primarily used in Europe. Efforts are being made for a universal acceptance of one bus system. At present, process control equipment are being manufactured for one system or the other; global acceptance of equipment standards would be preferred. A serial data bus is a single pair of twisted copper wires and enables communication between a central processing computer and many monitoring points and actuators when smart sensors are used. This is shown in Fig. 13.12. Although initially more expensive than direct lead connections, the advantages of the serial bus are minimal bus cost and installation labor. The system replaces all the leads to all the monitoring points by one pair of leads, new units can be added to the bus with no extra wiring, a plug and play feature is provided, giving faster control, and programming that is the same for all systems. Higher accuracies are obtained than using analog and more powerful diagnostics are available. As the cost of integration and development lowers, the bus system with its features will become more cost effective than the present systems.

The bus system uses time division multiplexing. The serial data word from the central processor contains the address of the peripheral unit being addressed in a given time slot and the data being sent. In the FF current from a constant current, supply is digitally modulated. Information on FF is given in the ISA 50.02 standards.

One disadvantage of the FF is that a failure of the bus, such as a broken wire, can shut the whole process down, whereas, with the direct connection method only one sensor is disabled. This disadvantage can be overcome by the use of a redundant or a backup bus in parallel to the first bus, so that if one bus malfunctions the backup bus can be used.

A comparison of the characteristics of the serial data buses is given in Table 13.1. The original FF was designated H1, a new generation of the H1 is the HSE, which will use an Ethernet LAN bus to provide operation under the TCP/IP protocol

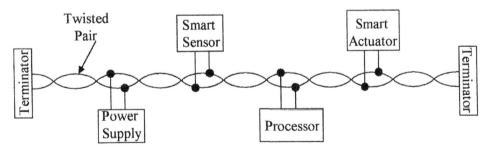

Figure 13.12 Foundation Fieldbus cable connection.

TABLE 13.1 Comparison of Bus Characteristics

	Fieldbus (H1)	Profibus	Fieldbus (HSE)
Bus type	twisted pair copper	twisted pair copper or fiber-optic cable	twisted pair copper and fiber-optic cable
Number of devices	240 per segment 65,000 segments	127 per segment 65,000 segments	Unlimited
Length	1900 m	100 m copper plus 24 km fiber	100 m copper plus 24 km fiber
Max speed	31.25 kb/s	12 Mb/s	100 Mb/s
Cycle time (millisecond)	<600	<2	<5

used for the internet. The advantages are increased speed, unlimited addresses, and standardization.

13.5 Controller

Due to the complexity and large number of variables in many process control systems, microprocessor based Programmable Logic Controllers (PLC) are used for decision making. The PLC can be configured to receive a small number of inputs (both analog and digital) and control a small number of outputs or the system can be expanded with plug-in modules to receive a large number of signals and simultaneously control a large number of actuators, displays, or other types of devices. In very complex systems, PLCs have the ability to communicate with each other on a global basis and to send operational data to and be controlled from a central computer terminal. Figure 13.13 shows a typical controller setup, monitoring a single variable. The output from the sensor is conditioned and transmitted to the input module of the controller; if the signal is an analog signal it is converted to a digital signal and compared to a reference signal stored in the computer. A decision can then be made and the appropriate control signal sent via the output module to the actuator.

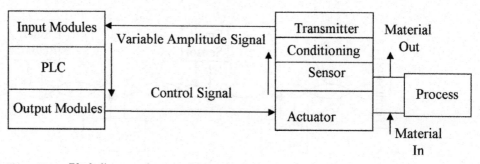

Figure 13.13 Block diagram of a control loop.

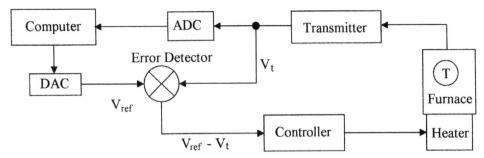

Figure 13.14 Supervisory control using an analog control loop.

The internal control function of a furnace using an analog loop is shown in Fig. 13.14. The temperature of the furnace is transmitted to the computer in the PLC as an analog signal, where it is converted to a digital signal and recorded in the computer memory. The voltage reference signal V_{ref} is converted to an analog signal and the two signals are subtracted as shown in an analog error differencing circuit. The amplified difference signal is then fed via a control circuit to the furnace heater, making the control loop analog.

In Fig. 13.15 the furnace temperature is converted into a digital signal and transmitted to the computer in the PLC. The digital temperature signal D_t and the digital reference signal D_{ref} are compared in a digital error detection circuit and the difference signal is sent as a digital signal to the controller, where it is converted into a *pulse width modulated* (PWM) signal or a DAC is used to control the heater. The control loop in this case would be considered as a digital control loop. PWM will be discussed later in this section.

13.5.1 Controller operation

The operation cycle in the PLC is made up of two separate modes; these are the I/O scan mode followed by the execution mode.

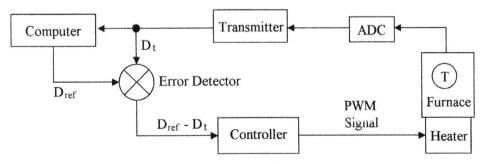

Figure 13.15 Supervisory control using a digital control loop.

I/O scan mode is the period when the processor updates the output control signals based on the information received from the previous I/O scan cycle after its evaluation of the signals. The processor then scans the inputs in a serial mode and updates its internal memory as to the status of the inputs.

Execution mode follows the I/O scan mode. In this mode the processor evaluates the input data stored in memory against the data programmed into the CPU. The programs are usually set up using ladder networks, where each rung of the ladder is an instruction for the action to be taken for each given input data level. The rung instructions are sequentially scanned and the input data evaluated. The processor can then determine the actions to be taken by the output modules and puts the data into memory for transfer to the output modules during the next I/O scan mode.

Scan time is the time required for the PLC to complete one I/O scan plus the execution cycle. This time depends on the number of input and output channels, the length of the ladder instruction sets, and the speed of the processor. A typical scan time is between 5 and 20 ms. As well as evaluating data, the PLC can also generate accurate time delays, store and record data for future use, and produce data in chart or graph form.

13.5.2 Ladder diagrams

The ladder diagram is universally used as a symbolic and schematic way to represent the interconnections between the elements in a PLC. The ladder network is also used as a tool for programming the operation of the PLC. The elements are interconnected as shown between the supply lines for each step in the control process, giving the appearance of the rungs in a ladder. A number of programming languages are in common use for controllers, they are as follows:

Ladder

Instruction list

Boolean flowcharts

Functional blocks

Sequential function charts

High level languages (ANSI, C, structured text)

Figure 13.16 shows some of the typical symbols used for the elements in a ladder diagram. A number of momentary action switches are shown, these are from top to bottom; a push to close (normally open NO) and a push to open (normally closed NC). These switches are the normal momentary action panel mounted operator switches.

Position limit switches are used to sense the position of an object and set to close or open when a desired position is reached. Pressure, temperature, and level switches are used to set limits and can be designed to open or close when the set limits are reached.

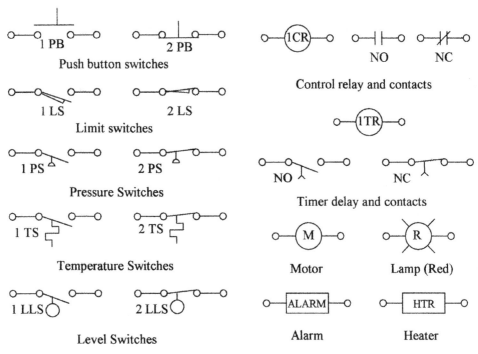

Figure 13.16 Some of the symbols in use for ladder diagrams.

The symbol for a control relay is a circle with the designation CR followed or preceded by a number to distinguish between the various relays used. This is shown with the symbols for its NO and NC contacts. These contacts will carry the same CR and number as the relay. A timer relay has the designation TR with a number and its associated NO and NC contacts will be likewise named and numbered.

A motor is represented by a circle with the letter M and an appropriate number. An indicator is represented by a circle with radiating arms and a letter to indicate its color, i.e., R = red, B = blue, O = Orange, G = green, and so on. Other elements are represented by boxes as shown, with the name of the element and a number to distinguish between similar types of elements used in different places.

The verticals forming the sides of the ladder represent the supply lines. The elements are connected serially between the supply lines as in a normal electrical schematic to form the rungs of the ladder. Each ladder rung is numbered using the hexadecimal numbering system with a note describing the function of the rung. This concept is best understood by Example 13.1.

Example 13.1 The heating system shown in Fig. 13.17a shows a container of liquid with a heating element, three momentary action push button switches (start, stop, alarm off) with red and green indicator lights, and an alarm. Show, using a ladder

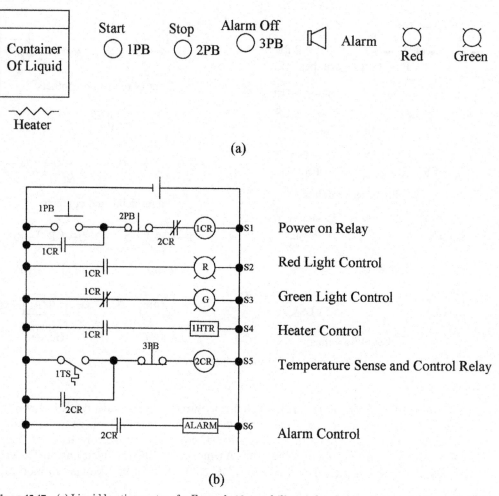

Figure 13.17 (a) Liquid heating system for Example 13.1 and (b) complete ladder diagram for Example 13.1.

diagram, how the elements can be connected, so that when system power is applied the green light is lit. When the start button is pushed the heater is turned ON, the red light turns on and green goes off. When the liquid reaches a preset temperature, the power is turned off and an alarm sounds. Provision is made to turn the alarm OFF and to be able to turn the power to the heater OFF.

Figure 13.17b shows a possible solution to Example 13.1. The first three rungs $S1$, $S2$, and $S3$ are used for control of the indicator lights and the power to the rest of the circuit. Momentary closure of switch 1 PB will energize control relay 1 CR, this in turn will close 1 CR NO contacts and open 1 CR NC contacts. These contacts will perform the following operations:

1. In ladder rung $S1$, 1 CR NO contact will connect 1 CR relay to the supply so that on release of 1 PB the relay will remain energized.
2. In ladder rung $S2$, the red light will be turned ON by the 1 CR NO contacts.
3. In ladder rung $S3$, the green light will be turned OFF by opening the 1 CR NC contacts.
4. In ladder lung $S4$, power will be turned ON to the heating element by the 1 CR contacts.

The container of liquid will heat up until the temperature sensor in rung $S5$ reaches its limit and closes. On closure, control relay 2 CR is energized and its contacts will perform the following functions:

1. In rung $S5$ the contact will connect 2 CR relay to the supply, bypassing the temperature sensor contacts, keeping 2 CR energized.
2. In rung $S6$ the contacts will supply power to the alarm.
3. In rung $S1$ the 2 CR NC contact will turn OFF the supply to control relay 1 CR de-energizing the relay, which in turn will change the lights from red to green and turn OFF the heating element.

The alarm will remain enabled until turned OFF by switch 3 PB via relay 2 CR. Relay 1 CR can also be turned OFF by switch 2 PB.

13.6 Digital-to-Analog Conversion

There are two methods of converting digital signals to analog signals. These are digital-to-analog converters, which are normally used to generate a voltage reference or low power voltage signals, and pulse width modulation that is used in high power circuits, i.e., actuator and motor control and so on.

13.6.1 Digital-to-analog converters

Digital-to-analog converters (DAC) change digital information into analog using a resistor network or similar method. The analog signals are normally used for low power applications but can be amplified and used for control. Figure 13.18a shows the generation of a 1 kHz sine wave. In the example shown, the digital signal is converted to a voltage every 0.042 ms giving the step waveform shown. In practice, the conversion rate could be higher approximating the steps to a complete sine wave. Shown also is the binary code from the DAC (4 bits only), the step waveform can be smoothed by a simple RC filter to get the sine wave. The example is only to give the basic conversion idea. Commercial DAC, such as the DAC 0808 shown in Fig. 13.18b are readily available. The DAC 0808 is an 8-bit converter which will give an output resolution or accuracy of 1 in $(2^8 - 1)$ (−1 is because the first number is zero leaving 255 steps) or an accuracy of ±0.39 percent. For higher accuracy analog signals a 12-bit commercial DAC would be used (±0.025 percent accuracy).

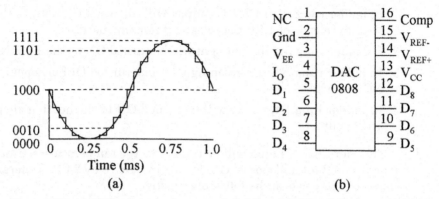

Figure 13.18 (a) 1 kHz sine waveform reproduced from a DAC and (b) commercial 8-bit DAC.

13.6.2 Pulse width modulation

Pulse Width Modulation changes the duration for which the voltage is applied to reproduce an analog signal and is shown in Fig. 13.19. The width of the output pulses shown are modulated, going from narrow to wide and back to narrow. If the voltage pulses shown are averaged, the width modulation shown will give a half-sine wave. The other half of the sine wave is generated using the same modulation, but with a negative supply, or with the use of a bridge circuit to reverse the current flow. The current is limited by the load. This type of width modulation is normally used for power drivers for ac motor control from a dc supply. Output devices such as the insulated gate bipolar transistor (IGBT) are used as switches, i.e., they are ON or OFF and can control over 100 kW of power. This method of conversion gives low internal dissipation with high efficiency,

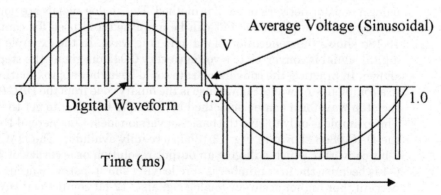

Figure 13.19 PWM signal to give a 1 kHz sine wave using positive and negative supplies.

that can be as high as 95 percent of the power going into the load. Whereas, analog power drivers are only 50 percent efficient at best and have high internal power dissipation.

13.7 Telemetry

Telemetry is the wireless transmission of measurement data from a remote location to a central location for processing and/or storage. This type of transmission is used for sending data over long distances from weather stations and the like, and data from rotating machinery where cabling is not feasible. More recently wireless communication is being used to eliminate cabling or to give flexibility in moving the positioning of temporary monitoring equipment. Broadcast information is a wireless transmission using amplitude modulation (AM) or frequency modulation (FM) techniques. But these methods are not accurate enough for the transmission of instrumentation data, as reception quality varies and the original signal can not be accurately reproduced. In telemetry, transmitters transmit signals over long distances using a form of FM or a variable width amplitude modulated signal. When transmitting from battery or solar cell operated equipment it is necessary to obtain the maximum transmitted power for the minimum power consumption. FM transmits signals at a constant power level, whereas, AM transmits at varying amplitudes and pulsing techniques can transmit only the pulse information needed which conserves battery power, hence for the transmission of telemetry data pulse AM is preferred.

13.7.1 Width modulation

Width coded signals or PWM are blocks of RF energy whose width is proportional to the amplitude of the instrumentation data. Upon reception the width can be accurately measured and the amplitude of the instrumentation signal reconstituted. Figure 13.20a shows the relation between the voltage amplitude of the instrumentation signal and the width of the transmitted pulses, when transmitting a series of 1 V signals and a series of voltages through 10 V.

For further power saving PWM can be modified to pulse position modulation (PPM). Figure 13.20b shows a typical PWM modulation and the equivalent PPM signal. The PWM signal shows an OFF period for synchronization of transmitter and receiver. The receiver then synchronizes on the rising edge of the transmitted zero; the first three pulses of the transmission are calibration pulses followed by a stream of width modulated data pulses. In the case of the PPM, narrow synchronization pulses are sent and then only a pulse corresponding to the lagging edge of the width modulated data is sent. Once synchronized, the receiver knows the position of the rising edge of the data pulses, so that information on the lagging edge is all that is required for the receiver to regenerate the data. This form of transmission has the advantage of greatly reducing power consumption and extending battery life.

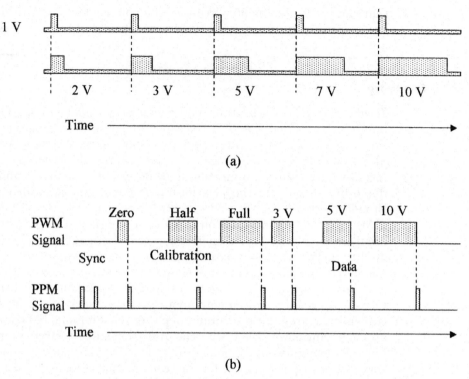

Figure 13.20 (*a*) an amplitude modulated waveform where the width of the modulations correspond to voltage levels and (*b*) PWM and PPM waveforms compared.

13.7.2 Frequency modulation

Frequency modulation is shown in Fig. 13.21a. The unmodulated carrier has fixed amplitude and frequency. In Fig. 13.21b the frequency of the transmitted signal is varied in proportion to the amplitude of the variable; the amplitude of the transmitted signal does not change. On reception the base frequency of the transmission is subtracted from the received signal leaving the frequency

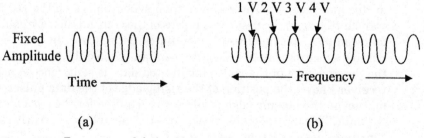

Figure 13.21 Frequency modulation (*a*) unmodulated waveform and (*b*) data modulated waveform.

of modulation, which can then be measured and the data reconstituted to determine the amplitude of the variable.

Summary

This chapter discusses the various types of data used in signal transmission and their advantages and disadvantages. The methods of interconnecting multiple devices in a process control system and controller operation.

The important points covered in this chapter are as follows:

1. The pros and cons of pneumatic signal transmission compared to electrical transmission in new and old systems.

2. Electrical analog signal transmission can use voltage or current signals and both have their advantages and disadvantages when trying to minimize errors.

3. Electrical signals can be converted to pneumatic signals for driving actuators for controlling hazardous material, where electrical sparks could ignite the material and so forth.

4. Thermocouples can be set up and interconnected to measure differential and average temperatures or connected in series to increase their sensitivity as in a thermopile.

5. RTDs can be interconnected using two, three, or four wires. Two wires are the lowest cost method, but are less accurate than the more expensive four wire connection.

6. Digital signal transmission has higher signal integrity and is faster than analog signal transmission and can use error correction codes to correct for any errors in transmission.

7. Smart sensors and the interconnection schemes using FF in the United States and Profibus in Europe.

8. Considerations when programming a PLC using ladder networks and the symbols used in the network.

9. The PLC cycle is divided into two modes of operation; the scan mode and the execution mode.

10. Conversion from digital signals to analog control signals using resistor networks and pulse width modulation.

11. Transmission of telemetry signals and methods of reducing the power required by using PPM.

Problems

13.1 Name various methods of data transmission.

13.2 What types of signal transmissions are used in telemetry?

13.3 What are the various types of connections used for RTD elements and what are the advantages and disadvantages of the various types of connections?

13.4 What are the standard ranges used in the transmission of pneumatic and electrical analog signals?

13.5 What conversion techniques are used to convert digital into analog signals?

13.6 Define offset and span.

13.7 What are the digital transmission standards?

13.8 Describe where a fiber cable is used and its advantages and disadvantages.

13.9 What are the advantages of digital over analog transmission?

13.10 Describe the modes of operation and scan time of a PLC.

13.11 What transmission speeds are used in the FF system?

13.12 What is a smart sensor?

13.13 Describe PPM and its advantage over PWM.

13.14 What is a ladder diagram?

13.15 What are the advantages of amplitude modulation over frequency modulation?

13.16 What are the advantages and disadvantages of current over voltage signal transmission?

13.17 Why are pneumatic signals used in electrical signal transmission?

13.18 Name the modes used in a controller.

13.19 How many steps are there in a 12-bit DAC and what is the percent resolution?

13.20 How many devices can be connected to the Fieldbus and Profibus?

Chapter 14

Process Control

Chapter Objectives

This chapter is an introduction to different process control concepts and will help you understand and become familiar with the different process control actions.
  The topics covered in this chapter are as follows:

- Concepts of signal control and controller modes
- Concept of lag time, error signals, and correction signals
- ON/OFF types of process controller action
- Proportional, derivative, and integral action in process controllers
- ON/OFF pneumatic control systems
- ON/OFF electric controllers
- Pneumatic proportional, integral, and derivative (PID) controllers
- Analog electronic implementation of proportional, derivative, and integral action
- An electronic PID loop
- Digital controller system

14.1 Introduction

Control systems vary extensively in complexity and industrial application. Industrial controllers, for instance, in the petrochemical industry, automotive industry, soda processing industry, and the like, have completely different types of control functions. The control loops can be very complex, requiring microprocessor supervision, down to very simple loops such as those used for controlling water temperature or heating, ventilation, and air-conditioning (HVAC) for comfort. Some of the functions need to be very tightly controlled, with tight tolerance on the variables and a quick response time, while in other areas the tolerances and response times are not so critical. These systems are closed loop systems. The output level is monitored against a set reference level and any difference detected between the two is amplified and used to control an input variable which will maintain the output at the set reference level.

14.2 Basic Terms

Some of these terms have already been defined, but apply to this chapter. Hence, the terms are redefined here for completeness.

Lag time is the time required for a control system to return a measured variable to its set point after there is a change in the measured variable, which could be the result of a loading change or set point change and so on.

Dead time is the elapse time between the instant an error occurs and when the corrective action first occurs.

Dead-band is a set hysteresis between detection points of the measured variable when it is going in a positive or a negative direction. This band is the separation between the turn ON set point and the turn OFF set point of the controller and is sometimes used to prevent rapid switching between the turn ON and turn OFF points.

Set point is the desired amplitude of an outpoint variable from a process.

Error signal is the difference between a set reference point and the amplitude of the measured variable.

Transient is a temporary variation of a load parameter after which the parameter returns to its nominal level.

Measured variable is an output process variable that must be held within given limits.

Controlled variable is an input variable to a process that is varied by a valve to keep the output variable (measured variable) within its set limits.

Variable range is the acceptable limits within which the measured variable must be held and can be expressed as a minimum and a maximum value, or a nominal value (set point) with ± spread (percent).

Control parameter range is the range of the controller output required to control the input variable to keep the measured variable within its acceptable range.

Offset is the difference between the measured variable and the set point after a new controlled variable level has been reached. It is that portion of the error signal which is amplified to produce the new correction signal and produces an "Offset" in the measured variable.

14.3 Control Modes

The two basic modes of process control are ON/OFF action and "continuous control" action. In either case the purpose of the control is to hold the measured variable output from a process within set limits by varying the controlled input variable to the process.

In the case of ON/OFF control (discrete control or two position control), the output of the controller changes from one fixed condition (ON) to another fixed position (OFF). Control adjustments are the set point and in some applications a dead-band is used. In continuous control (modulating control) action the feedback controller determines the error between a set point and a measured variable. The error signal is then used to produce an actuator control signal to operate a valve and reduce the error signal. This type of control continuously monitors the measured variable and has three modes of operation which are proportional, integral, and derivative. Controllers can use one of the functions, two, or all three of the functions as required.

14.3.1 ON/OFF action

The simplest form of control in a closed loop system is ON/OFF action. The measured variable is compared to a set reference. When the variable is above the reference the system is turned ON and when below the reference the system is turned OFF or vice versa, depending upon the system design. This could make for rapid changes in switching between states. However, such systems normally have a great deal of inertia or momentum which produces overswings and introduces long delays or lag times before the variable again reaches the reference level. Figure 14.1a shows an example of a simple room heating system. The top graph shows the room temperature or measured variable and the lower graph shows the actuator signal. The room temperature reference is set at 75°F. When the air is being heated, the temperature in the center of the room has already reached 77°F before the temperature at the sensor reaches the reference temperature of 75°F

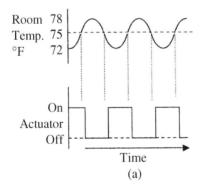 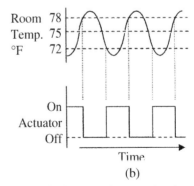

Figure 14.1 A room heating system with (a) simple ON/OFF action of a room heating system and (b) differential ON/OFF action.

and similarly as the room cools, the temperature in the room will drop to 73°F before the temperature at the sensor reaches 75°F. Hence, the room temperature will go from about 72°F to 78°F due to the inertia in the system.

14.3.2 Differential action

Differential or delayed ON/OFF action is a mode of operation where the simple ON/OFF action has hysteresis or a dead-band built in. Figure 14.1b shows an example of a room heating system similar to that shown in Fig. 14.1a except that instead of the thermostat turning ON and OFF at the set reference of 75°, the switching points are delayed by ±3°F. As can be seen in the top graph, the room temperature reaches 78°F before the thermostat turns OFF the actuator and the room temperature falls to 72°F before the actuator is turned ON giving a built in hysteresis of 6°F. There is, of course, still some inertia. Hence, the room temperature will go from about 70°F to about 80°F.

14.3.3 Proportional action

The most common of all continuous industrial process control action is proportional control action. The amplitude of the output variable from a process is measured and converted to an electrical signal. This signal is compared to a set reference point. Any difference in amplitude between the two (error signal) is amplified and fed to a control valve (actuator) as a correction signal. The control valve controls one of the inputs to the process. Changing this input will result in the output amplitude changing until it is equal to the set reference or the error signal is zero. The amplitude of the correction signal is transmitted to the actuator controlling the input variable and is proportional to the percentage change in the output variable amplitude measured with respect to the set reference. In industrial processing a different situation exists than with a room heating system. The industrial system has low inertia; overshoot and response times must be minimized for fast recovery and to keep processing tolerances within tight limits. In order to achieve these goals fast reaction and settling times are needed. There may also be more than one variable to be controlled and more than one output being measured in a process.

The change in output level may be a gradual change, a large on-demand change, or caused by a change in the reference level setting. An example of an on-demand change would be cleaning stations using hot water at a required fixed temperature, as shown in Fig. 14.2a. At one point in time the demand could be very low with a low flow rate as would be the case if only one cleaning station were in use. If cleaning commenced at several of the other stations, the demand could increase in steps or there could be a sudden rise to a very high flow rate. The increased flow rate would cause the water temperature to drop. The drop in water temperature would cause the temperature sensor to send a correction signal to the actuator controlling the steam flow, so as to increase the steam flow to raise the temperature of the water to bring it back to the set reference level (see Fig. 14.2b). The rate of correction will depend on the inertia in the system, gain in the feedback loop, allowable amount of overshoot, and so forth.

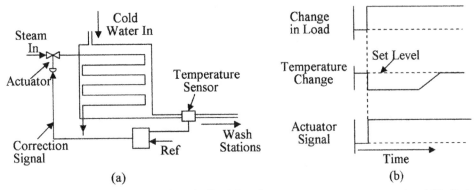

Figure 14.2 Water heater (a) showing a feedback loop for constant temperature output and (b) effect of load changes on the temperature of the water from the water heater.

In a closed loop feedback system settings are critical. If the system has too much gain, i.e., the amplitude of the correction signal is too great, it will cause the controlled variable to over correct for the error, which in turn will give a false error signal in the reverse direction. The actuator will then try to correct for the false error signal. This can, in turn, send a larger correction signal to the actuator, which will cause the system to oscillate or cause an excessively long settling or lag time. If the gain in the system is too low the correction signal is too small and the correction will never be fully completed, or again an excessive amount of time is taken for the output to reach the set reference level.

This effect is shown in Fig. 14.3 (a), as can be seen in comparing the over corrected (excessive gain) and the under corrected (too little gain) to the optimum

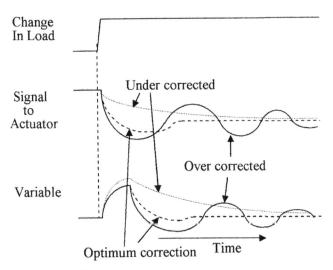

Figure 14.3 Effect of loop gain on correction time using proportional action with over correction and under correction.

gain case (with just a little overshoot). The variable takes a much longer time for the correction to be implemented than in the optimum case. In many processes this long delay or lag time is unacceptable.

14.3.4 Derivative action

Proportional plus derivative (PD) action was developed in an attempt to reduce the correction time that would have occurred using proportional action alone. Derivative action senses the rate of change of the measured variable and applies a correction signal that is proportional to the rate of change only (this is also called rate action or anticipatory action). Figure 14.4a shows some examples of derivative action. As can be seen in this example, a derivative output is obtained only when the load is changing. The derivative of a positive slope is a positive signal and the derivative of a negative slope is a negative signal; zero slopes give zero signals as shown. An in-depth look at derivatives is outside the scope of this text.

Figure 14.4b shows the effect of PD action on the correction time. When a change in loading is sensed as shown, both the proportional and derivative signals are generated and added. The significance of combining these two signals is to produce a signal that speeds up the actuator's control signal. The faster reaction time of the control signal reduces the time to implement corrective action reducing the excursion of the measured variable and its settling time. The amplitudes of these signals must be adjusted for optimum operation, or overshoot or under shoot can still occur.

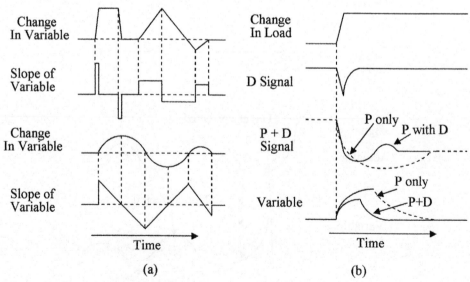

Figure 14.4 Proportional and derivative action (*a*) variable change with resulting slope and (*b*) effect of proportional and derivative action on a variable.

14.3.5 Integral action

Proportional plus Integral (PI) action also known as reset action, was developed to correct for long-term loads and applies a correction proportional to the area under the change in the variable curve. Figure 14.5a gives some examples of the integration of a curve or the area under a curve. In the top example the area under the square wave increases rapidly but remains constant when the square wave drops back to zero. In the triangular section the area increases rapidly at the apex but increases slowly as the triangle approaches zero; when the triangle goes negative, the area reduces. In the lower example the area increases more rapidly when the sine wave is at its maximum and slower as it approaches the zero level. During the negative portion of the sine wave, the area is reduced. Proportional action gives a response to a change in the measured variable but does not fully correct the change in the measured variable due to its limited gain. For instance, if the gain in the proportional amplifier is 100, then when a change in load occurs, 99 percent of the change is corrected. However, a 1 percent error signal is required for amplification to drive the actuator to change the manipulated variable. The 1 percent error signal is effectively an "offset" in the variable with respect to the reference. Integral action gives a slower response to changes in the measured variable to avoid overshoot, but has a high gain so that with long-term load changes it takes over control of the manipulated variable and applies the correction signal to the actuator. Because of the

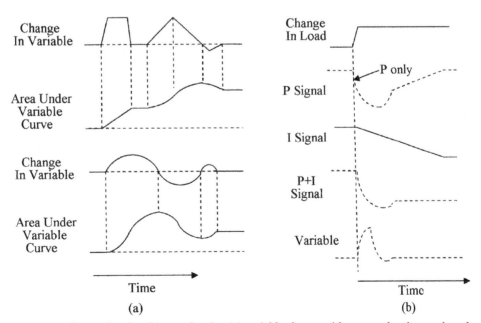

Figure 14.5 Proportional and integral action (*a*) variable change with area under the graph and (*b*) effect of proportional and integral action on a variable.

higher gain the measured variable error is reduced to close to zero. This also returns the proportional amplifier to its normal operating point, so that it can correct for other fluctuations in the measured variable. Note that these corrections are done at relatively high speeds. The older pneumatic systems are much slower and can take several seconds to make such a correction. Figure 14.5b shows the PI corrective action waveforms. When a change in loading occurs, the P signal responds to take corrective action to restore the measured variable to its set point; simultaneously, the integral signal starts to change linearly to supply the long term correction, thus allowing the proportional signal to return to its normal operating point as is shown. Here again integral action can become complex and further discussion is considered to be outside the scope of this text.

14.3.6 PID action

A combination of all three of the actions described above is more commonly referred to as PID action. The waveforms of PID action are illustrated in Fig. 14.6. PID is the most often used corrective action for process control. There are however, many other types of control actions based upon PID action. Understanding the fundamentals of PID action gives a good foundation for understanding other

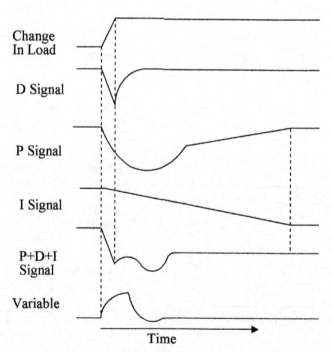

Figure 14.6 Waveforms for proportional plus integral action and waveforms for proportional plus derivative and integral action.

types of controllers. The waveforms used have been idealized for ease of the explanation and are only an example of what may be encountered in practice. Loading is a function of demand and is not affected by the control functions or actions; the control function is to ensure that the variables are within their specified limits.

To give an approximate indication of the use of PID controllers for different types of loops, the following are general rules that should be followed:

Pressure control requires proportional and integral; derivative is normally not required.

Level control uses proportional and sometimes integral, derivative is not normally required.

Flow control requires proportional and integral; derivative is not normally required.

Temperature control uses proportional, integral, and derivative usually with integral set for a long time period.

However, the above are general rules and each application has its own requirements.

Typical feedback loops have been discussed. The reader should, however, be aware that there are other kinds of control loops used in process control such as cascade, ratio, and feed-forward.

14.4 Implementation of Control Loops

Implementation of the control loops can be achieved using pneumatic, analog, or digital electronics. The first process controllers were pneumatic. However, these have largely been replaced by electronic systems, because of improved reliability, less maintenance, easier installation, easier adjustment, higher accuracy, lower cost, can be used with multiple variables, and have higher speed operation.

14.4.1 ON/OFF action pneumatic controller

Figure 14.7 shows a pneumatic furnace control system using a pneumatic ON/OFF controller. In this case the furnace temperature sensor moves a flapper that controls the air flow from a nozzle. When the temperature in the furnace reaches its set point the sensor moves the flapper toward the nozzle to stop the air flow and allow pressure to build up in the bellows. The bellows operates an air control relay that shuts OFF the air flowing to the control valve turning OFF the fuel to the furnace. When the temperature in the furnace drops below a set level the flapper is opened by the sensor, reducing the air pressure in the bellows, which in turn opens the air control valve allowing the air pressure to drop and the control valve to open, turning ON the fuel to the furnace.

250 Chapter Fourteen

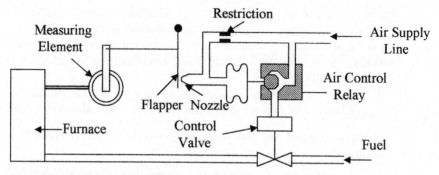

Figure 14.7 Pneumatic ON/OFF furnace controller.

14.4.2 ON/OFF action electrical controller

An example of an ON/OFF action electrical room temperature controller is shown in Fig. 14.8. In this case the room temperature is sensed by a bimetallic sensor. The sensor operates a mercury switch. As the temperature decreases the bimetallic element tilts the mercury switch down causing the mercury to flow to the end of the glass envelope and in so doing shorts the two contacts together in the mercury switch. The contact closure operates a low voltage relay turning ON the blower motor and the heating element. When the room temperature rises to a predetermined set point the bimetallic strip tilts the mercury switch back causing the mercury to flow away from the contacts. The low voltage electrical circuit is turned OFF, the relay opens, and the power to the heater and the blower motor is disconnected.

The ON/OFF controller action has many applications in industry; an example of some of these uses is shown in Fig. 14.9. In this case, cartons on a conveyer belt are being filled from a hopper. When a carton is full it is sensed by the level sensor, which sends a signal to the controller to turn OFF the material flowing from the hopper and to start the conveyer moving. As the next carton moves into the filling position it is sensed by the position sensor, which

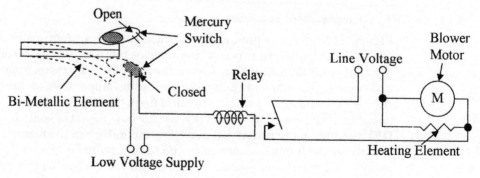

Figure 14.8 Simple ON/OFF room heating controller.

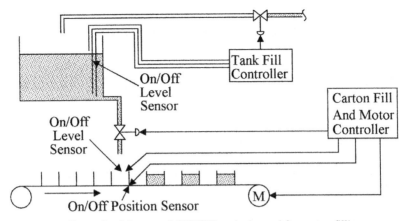

Figure 14.9 Example of the use of ON/OFF controls used for carton filling.

sends a signal to the controller to stop the conveyer belt and to start filling the carton. Once it is full the cycle repeats itself.

A level sensor in the hopper senses when the hopper is full and when it is almost empty. When empty, the sensor sends a signal to the controller to turn ON the feed valve to the hopper and when the hopper is full it is detected and a signal is sent to the controller to turn the feed to the hopper OFF.

14.4.3 PID action pneumatic controller

Many configurations for PID pneumatic controllers have been developed over the years, have served us well, and are still in use in some older processing plants. But pneumatic controllers have, with the advent of the requirements of modern processing and the development of electronic controllers, achieved the distinction of becoming museum pieces. Figure 14.10 shows an example of a pneumatic PID

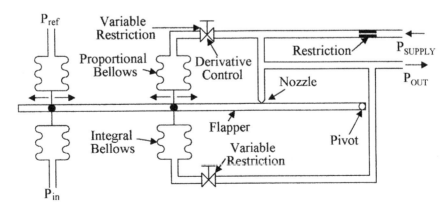

Figure 14.10 Pneumatic PID controller.

controller. The pressure from the sensing device P_{in} is compared to a set or reference pressure P_{ref} to generate a differential force (error signal) on the flapper to move the flapper in relation to the nozzle giving an output pressure proportional to the difference between P_{in} and P_{ref}. If the derivative restriction is removed the output pressure is fed back to the flapper via the proportional bellows to oppose the error signal and to give proportional action. System gain is adjusted by moving the position of the bellows along the flapper arm, i.e., the closer the bellows is positioned to the pivot the greater the movement of the flapper arm.

By putting a variable restriction between the pressure supply and the proportional bellows, a change in P_{in} causes a large change in P_{out}, as the feedback from the proportional bellows is delayed by the derivative restriction. This gives a pressure transient on P_{out} before the proportional bellows can react, thus giving derivative action. The duration of the transient is set by the size of the bellows and the setting of the restriction.

Integral action is achieved by the addition of the integral bellows and restriction as shown. An increase in P_{in} moves the flapper towards the nozzle causing an increase in output pressure. The increase in output pressure is fed to the integral bellows via the restriction until the pressure in the integral bellows is sufficient to hold the flapper in the position set by the increase in P_{in}, creating integral action.

14.4.4 PID action control circuits

PID action can be performed using either analog or digital electronic circuits. In order to understand how electronic circuits are used to perform these functions, the analog circuits used for the individual actions will be discussed. The circuit shown in Fig. 14.11a is used to compare the signal from the measured variable and the reference to generate the error signal. Proportional action is achieved as shown in Fig. 14.11b by amplifying the error signal V_{in}. The stage

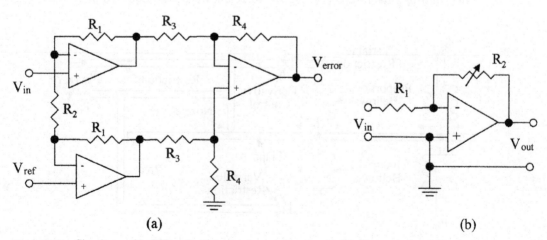

Figure 14.11 Circuits used in PID action (a) error generating circuit and (b) proportional circuit.

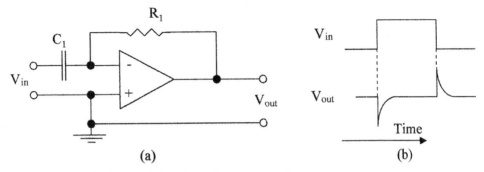

Figure 14.12 Derivative amplifier (a) circuit and (b) waveforms.

gain is the ratio of R_2/R_1; the gain can be adjusted using the potentiometer R_2. The output is inverted.

The circuit for derivative action is shown in Fig. 14.12a. The feedback resistor can be replaced with a potentiometer to adjust the differentiation duration. The output signal is inverted. This signal can be changed to a noninverted signal with an inverting amplifier stage if required. The waveforms of the differentiator are shown in Fig. 14.12b.

Proportional and derivative action can be combined using the circuit shown in Fig. 14.13a. Derivative action is obtained by the input capacitor C_1 and proportional action by the ratio of the resistors R_1 and R_2. The inverted output signal is shown in Fig. 14.3b.

A circuit to perform integral action is shown in Fig. 14.14a. Capacitive feedback around the amplifier prevents the output from the amplifier from following the input change. The output changes slowly and linearly when there is a change in the measured variable as shown in the waveforms in Fig. 14.14b. The slope of the output waveform is set by the time constant of the feedback C_1 and the input resistance R_1. This is integral action and the output from the integrator

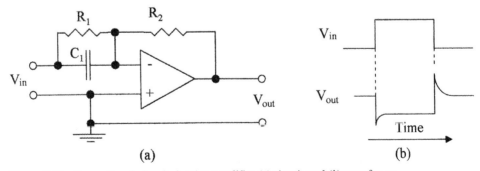

Figure 14.13 Proportional plus derivative amplifier (a) circuit and (b) waveforms.

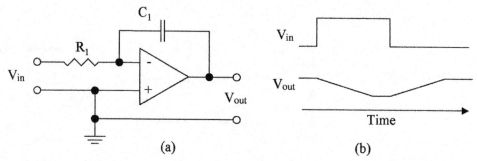

Figure 14.14 Integrating amplifier (*a*) circuit and (*b*) waveforms.

is the area under the input waveform. This area can be adjusted by replacing R_1 with a potentiometer. The output of the amplifier is inverted.

14.4.5 PID electronic controller

Figure 14.15 shows the block diagram of an analog PID controller. The measured variable from the sensor is compared to the set point in the first unity gain comparator; its output is the difference between the two signals or the error signal. This signal is fed to the integrator via an inverting unity gain buffer and to the proportional amplifier and differentiator via a second inverting unity gain comparator, which compares the error signal to the integrator output. Initially, with no error signal the output of the integrator is zero so that the zero error signal is also present at the output of the second comparator.

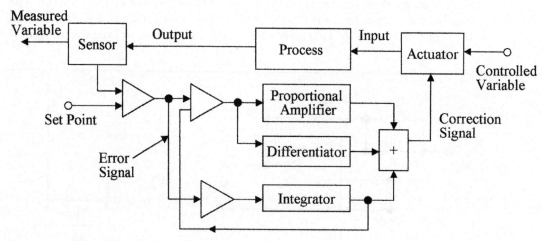

Figure 14.15 Block schematic of a PID electronic controller.

When there is a change in the measured variable, the error signal is passed through the second comparator to the proportional amplifier and the differentiator where it is amplified in the proportional amplifier, added to the differential signal in a summing circuit, and fed to the actuator to change the input variable. Although the integrator sees the error signal, it is slow to react and so its output does not change immediately, but starts to integrate the error signal. If the error signal is present for an extended period of time, the integrator will supply the correction signal via the summing circuit to the actuator and input the correction signal to the second comparator. This will reduce the effective error signal to the proportional amplifier to zero, when the integrator is supplying the full correction signal to the actuator. Any new change in the error signal will still be passed through the second comparator as the integrator is only supplying an offset to correct for the first long-term error signal. The proportional and differential amplifiers can then correct for any new changes in the error signal.

The circuit implementation of the PID controller is shown in Fig. 14.16. This is a complex circuit because all the amplifier blocks are shown doing a single function to give a direct comparison to the block diagram and is only used as an example. In practice there are a large number of circuit component combinations that can be used to produce PID action.

A single amplifier can also be used to perform several functions which would greatly reduce the circuit complexity. Such a circuit is shown in Fig. 14.17, where feedback from the actuator position is used as the proportional band adjustment.

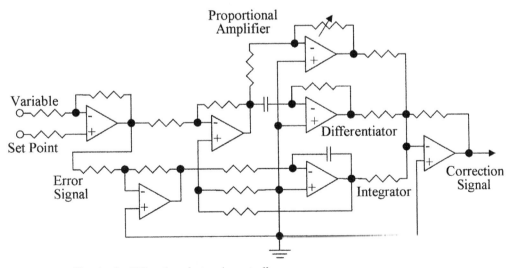

Figure 14.16 Circuit of a PID action electronic controller.

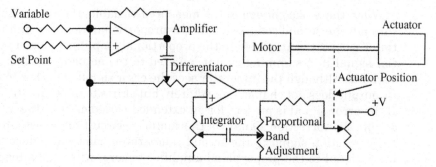

Figure 14.17 Circuit of a PID electronic controller with feed back from the actuator position.

In new designs, PLC processors can be used to replace the analog circuits to perform the PID functions using digital techniques.

14.5 Digital Controllers

Modern process facilities will use a computer or PLC processor as the heart of the control system. The system will be able to control analog loops, digital loops, and will have a foundation fieldbus input/output for communication with smart sensors. All of these control functions may not be required in small process facilities but in large facilities they are necessary. The individual control loops are not independent in a process but are interrelated and many measured variables may be monitored and manipulated variables controlled simultaneously. Several processors may also be connected to a mainframe computer for complex control functions. Figure 14.18 shows the block diagram of

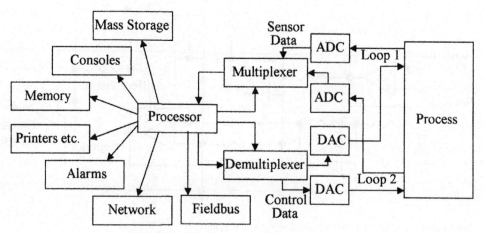

Figure 14.18 Computer based digital controlled process.

a processor controlling two digital loops. The analog output from the monitors is converted to a digital signal in an ADC. The loop signal is then selected in a multiplexer and PID action is performed in the processor using software programs. The digital output signal is then fed to the actuator through a demultiplexer and a DAC. The processor has mass storage for storing process data for later use or making charts and graphs and will also be able to control a number of peripheral units and monitors as shown.

Digital controllers will compare the digitized measured variable to the set point stored in memory to produce an error signal which it can amplify under program control and feed to an actuator via a DAC. The processor can measure the rate of change of the measured variable and produce a differential signal to add to the digital correction signal. In addition, the processor can measure the area under the measured variable signal which it will also add to the digital correction signal. All of these actions are under program control; the setting of the program parameters can be changed with a few key strokes, making the system much more versatile than the analog equivalent.

Summary

This chapter discussed process control and the various methods of implementation of the controller functions. Various controller modes and the methods of implementing the modes in pneumatic and electronic circuits are described. Understanding of these circuits will enable the reader to extend these principles to other methods of control.

The salient points covered in this section were:

1. ON/OFF and delayed ON/OFF action and their use in HVAC. A number of examples of ON/OFF action in process control were given.

2. Proportional, integral, and derivative action and their use in process control. The effects of gain setting in proportional control.

3. Circuits to perform proportional, integral, differential action and methods of combining the various actions in a PID controller are described.

4. The operation of pneumatic controller actions using flappers, nozzles, and bellows combinations is given. A combination of the various pneumatic components is used to make a PID controller.

5. Digital controller concepts in modern processing facilities are given.

Problems

14.1 Describe controller ON/OFF action.

14.2 What is the difference between simple ON/OFF action and differential ON/OFF action?

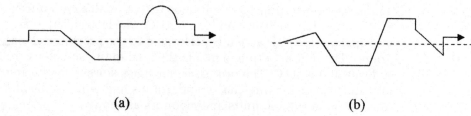

Figure 14.19 Change in measured variable for Prob. 14.6 through 14.9.

14.3 What is proportional action?

14.4 What is integral action?

14.5 What is derivative action?

14.6 Draw the derivative signal for the variable shown in Fig. 14.19*a*.

14.7 Draw the integral signal for the variable shown in Fig. 14.19*a*.

14.8 Draw the derivative signal for the variable shown in Fig. 14.19*b*.

14.9 Draw the integral signal for the variable shown in Fig. 14.19*b*.

14.10 Redraw the PID action controller in Fig. 14.16 as it would be if integral action was not required.

14.11 Give a list of applications for ON/OFF controller action.

14.12 Why is the gain setting critical in proportional action?

14.13 What is the difference between an error signal and a measured variable signal?

14.14 What is the difference between lag time and dead time?

14.15 What is the difference between offset and error signal?

14.16 What are some of the actions that can be taken to reduce correction time?

14.17 What is a dead-band?

14.18 What would be the effect of time constants on correction time?

14.19 What types of control do not normally require derivative action?

14.20 Why is ON/OFF action not normally suitable for control of a process?

Chapter 15

Documentation and Symbols

Chapter Objectives

This chapter will help you understand the need for documentation and become familiar with the use of symbols in process control flow diagrams.

The following are covered in this chapter:

- Alarm and trip systems and documentation
- Programmable logic controller (PLC) documentation
- Interconnection symbols and flow line abbreviations used in piping diagrams
- Instrument Society of America (ISA) list of standard symbols
- Standard instrument symbols and identification letters
- Standard functional actuator symbols
- Control loop numbering system and pipe and identification (P and ID) drawings

15.1 Introduction

Documentation covers front-end engineering and detailed engineering drawings. Of the overwhelming amount of documentation needed in a plant, the only documentation that will be introduced is limited to documentation you may encounter and need to use, such as alarm and trip systems, PLC documentation, and pipe and identification diagrams (P and ID). Of these the P and ID is the detailed documentation covering instruments, their location, process control loops, and process flow details. Documentation standards and symbols have been set up and standardized by the ISA in conjunction with the American National Standards Institute (ANSI).

15.2 System Documentation

15.2.1 Alarm and trip systems

Alarm and trip system information and implementation is given in ANSI/ISA-84.01-1996—Application of Safety Instrumented Systems for the Process Control Industry. The purpose of an alarm system is to bring a malfunction to the attention of operators and maintenance personnel, whereas that of a trip system is to shut down a system in an orderly fashion when a malfunction occurs, or switch failed units over to standby units. The elements used in the process control system are the first line of warning of a failure. The sensors and instruments used in the alarm and trip system are the second line of defense and must be totally separate from those used in the process control system.

The alarm and trip system or Safety Instrumented System (SIS) has its own sensors, logic, and control elements so that under failure conditions it will take the process to a safe state to protect the personnel, facility, and environment. To ensure full functionality of the SIS, it must be regularly tested. In the extreme with deadly chemicals, a second SIS system and redundancy can be used in conjunction with the first SIS system to ensure 100 percent protection. The sensors in the SIS will usually be of a different type than those used for process control. The control devices are used to accurately sense varying levels in the measured variable, whereas the SIS sensor is used to sense a trip point and can be a much more reliable rugged device. The use of redundancy in a system must not be used as a justification for low reliability cheap components. A common SIS system is the dual redundancy system that consists of the main SIS with two redundant systems. In this case, a two out of three logic monitoring system determines if a monitor has failed or the system has failed by correlation between the outputs. A two out of three logic circuit is shown in Fig. 15.1a. The truth table is shown in Fig. 15.1b; the inputs are normally low (0). If one input goes high (1) it would indicate a monitor failure and the monitor failure output would go from 0 to 1 but the system output would remain at 0. If 2 or more inputs go high it would indicate a system failure and the system failure output would go from 0 to 1 as shown.

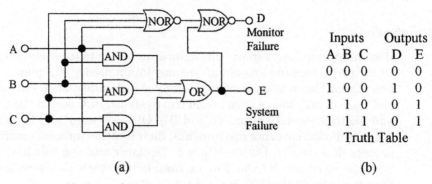

Figure 15.1 Monitor and (a) two out of three failure indicator and (b) truth table.

In SIS systems failure analysis, the rate of component failure is as follows:

Logic	8%
Sensors	42%
Control devices	50%

15.2.2 Alarm and trip documentation

Good up-to-date documentation is a must in alarm and trip systems. All SIS devices should be clearly marked and numbered. System drawings must show all SIS devices using standard symbols, their location, function, and set limits. Drawings must include lock and logic diagrams.

The types of information required in alarm and trip documentation are as follows:

1. Safety requirement specifications
2. Logic diagram with functional description
3. Functional test procedures and required maintenance
4. Process monitoring points and trip levels
5. Description of SIS action if tripped
6. Action to be taken if SIS power is lost
7. Manual shut down procedures
8. Restarting procedures after SIS shut down

15.2.3 PLC documentation

As with all technical devices detailed engineering records are essential. Without accurate drawings, changes and modifications needed for upgrading are extremely difficult or impossible. Every wire from the PLC to the monitoring and control equipment must be clearly marked at both ends and shown on the wiring diagram to facilitate wiring changes and diagnostics. The PLC must have complete up-to-date ladder diagrams (or other approved language). Every rung must be labeled with a complete description of its function.

The essential documents in a PLC package are as follows:

1. System overview and complete description of control operation
2. Block diagram of the units in the system
3. Complete list of every input and output, destination, and number
4. Wiring diagram of I/O modules, address identification for each I/O point, and rack location
5. Rung description, number, and function

15.3 Pipe and Identification Diagrams

15.3.1 Standardization

The electronics industry has standard symbols to represent circuit components for use in circuit schematics and similarly the processing industry has developed standard symbols to represent the elements in a process control system. Instead of a circuit schematic the processing industrial drawings are known as *pipe and identification diagrams* (P and ID) (not to be confused with PID) and represents how the components and elements in the processing plant are interconnected. Symbols have been developed to represent all of the components used in industrial processing and have been standardized by ANSI and ISA. The P and ID document is the ANSI/ISA S5.1–1984 (R 1992)—Instrumentation Symbols and Identification Standards. An overview of the symbols used is given in this chapter but the list is not complete. The ISA should be contacted for a complete list of standard symbols.

P and IDs or engineering flow diagrams were developed for the detailed design of the processing plant. The diagrams show complete details of all the required piping, instruments and location, signal lines, control loops, control systems, and equipment in the facility. The process flow diagrams and plant control requirements are generated by a team from process engineering and control engineering. Changes to the P and ID are normally the responsibility of process engineering and must be approved and signed off by the same. These engineering drawings must be correct, current, up-to-date, and rigorously maintained. Every P and ID change must be approved and recorded. If not, time is lost in maintenance, repair, and modifications, not to mention the catastrophic errors that can be made by using obsolete drawings.

P and ID typically show the following types of information:

1. Plant equipment and vessels showing location, capacity, pressure, liquid level operating range, usage and so on
2. All interconnection lines distinguishing between the types of interconnection, i.e., gas or electrical and operating range of line
3. All motors giving voltage and power and other relevant information
4. Instrumentation showing location of instrument, its major function, process control loop number, and range
5. Control valves giving type of control, type of valve, type of valve action, fail save features, and flow plus pressure information
6. The ranges for all safety valves, pressure regulators, temperatures, and operating ranges
7. All sensing devices, recorders, and transmitters with control loop numbers

15.3.2 Interconnections

The standard on interconnections specifies the type of symbols to be used to represent the various types of connections in a processing plant (see Fig. 15.2).

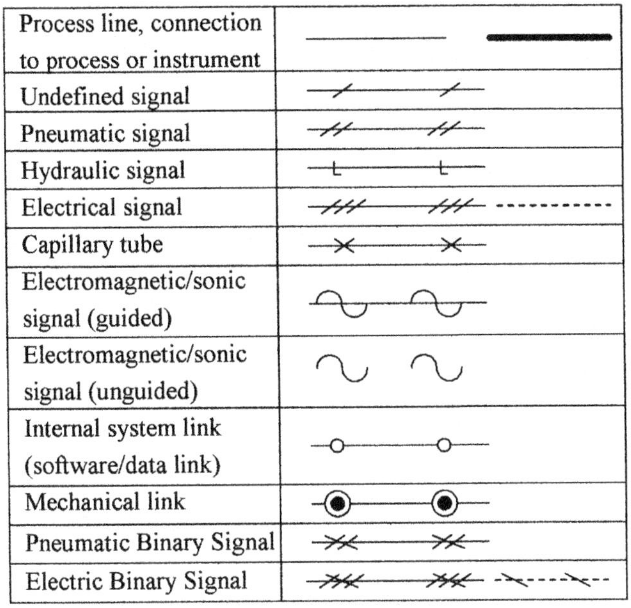

Process line, connection to process or instrument	——— ━━━
Undefined signal	—/— —/—
Pneumatic signal	—//— —//—
Hydraulic signal	—L— —L—
Electrical signal	—///— —///— - - - - -
Capillary tube	—✕— —✕—
Electromagnetic/sonic signal (guided)	∽∽
Electromagnetic/sonic signal (unguided)	∽ ∽
Internal system link (software/data link)	—o——o—
Mechanical link	—●——●—
Pneumatic Binary Signal	—⫽— —⫽—
Electric Binary Signal	—⫽— —⫽— ⤳- - -⤳- -

Figure 15.2 Symbols for instrument line interconnection.

The solid bold lines are used to represent the primary lines used for process product flow and the plain solid lines are used to represent secondary flows such as steam for heating. Abbreviations for secondary flow lines are given in Table 15.1. The abbreviations are placed adjacent to the lines to indicate their function as shown in Fig. 15.3.

In the list of assigned symbols for interconnect lines given in Fig. 15.2, one symbol is undefined and can be assigned at the users discretion for a special connection not covered by any of the assigned interconnection symbols. The binary signals can be used for digital signals or pulses. It is also necessary to show on the P and ID the signal's content and range. For example, electrical interconnections can be either signal current or voltage and would be marked as 4 to 20 mA or 0 to 5 V, examples of signal lines with the signals content and range marking are shown in Fig. 15.3.

15.3.3 Instrument symbols

Figure 15.4 shows the symbols designated for instruments. Discrete instruments are represented by circles, shared instruments by a circle in a rectangle,

TABLE 15.1 Abbreviations for Secondary Flow Lines

AS	Air supply	ES	Electric supply	GS	Gas supply
HS	Hydraulic supply	NS	Nitrogen supply	SS	Steam supply
WS	Water supply				

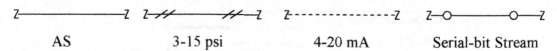

Figure 15.3 Method of indicating the signal content of a line.

computer functions by hexagons, and PLC functions by a diamond in a rectangle. A single horizontal line, no line, dashed line, or double line through the display is used to differentiate between location and accessibility to an operator, i.e., a line through an instrument may indicate the instrument is in a panel in the control room giving full access, no line could mean the instrument is in the process area and off limits to the operator, a double line means the possibility that the instrument is in a remote location but the operator can obtain access, whereas a dashed line means not available by virtue of being located in a totally inaccessible location.

15.3.4 Instrument identification

Instrument symbols should also contain letters and numbers. The letters are a shorthand way of giving the type of instrument, its use in the system, and the numbers identify the control loop. Usually 2 or 3 letters are used. The first letter identifies the measured or initiating variable, the following is a modifier, and the remaining letters identify the function. Table 15.2 shows some of the meaning of the assigned instrument letters.

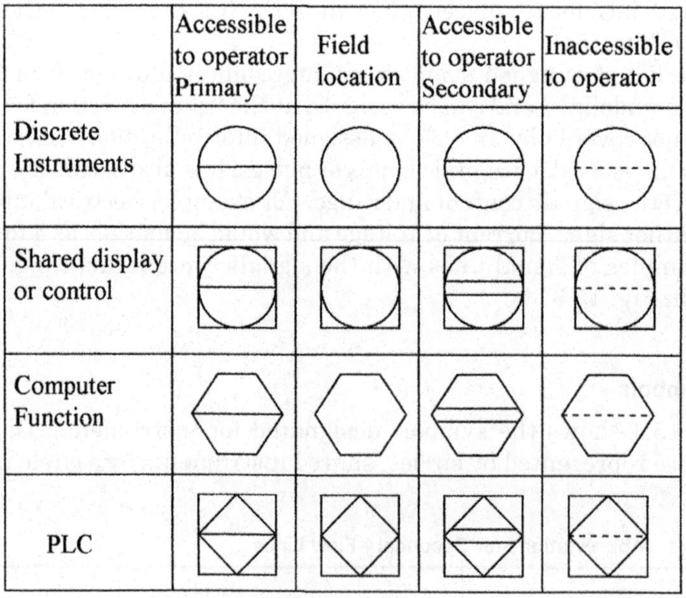

	Accessible to operator Primary	Field location	Accessible to operator Secondary	Inaccessible to operator
Discrete Instruments	⊖	○	⊖	(⊖ dashed)
Shared display or control	⊟	▢	⊟	(⊟ dashed)
Computer Function	⬡	⬡	⬡	(⬡ dashed)
PLC	◇	◇	◇	(◇ dashed)

Figure 15.4 Standardized instrument symbols.

TABLE 15.2 Instrument Identification Letters

	First letter + Modifier		Succeeding letters		
	Initiating or measured variable	Modifier	Readout or passive function	Output function	Modifier
A	Analysis		Alarm		
B	Burner, combustion		User's choice	User's choice	User's choice
C	User's choice			Control	
D	User's choice	Differential			
E	Voltage		Sensor		
F	Flow rate	Ratio			
G	User's choice		Glass, viewing device		
H	Hand				High
I	Current		Indicate		
J	Power	Scan			
K	Time	Time rate of change		Control station	
L	Level		Light		Low
M	User's choice	Momentary			Middle
N	User's choice		User's choice	User's choice	User's choice
O	User's choice		Orifice		
P	Pressure		Test point		
Q	Quantity	Integrate, totalize			
R	Radiation		Record		
S	Speed, frequency	Safety		Switch	
T	Temperature			Transmit	
U	Multivariable		Multifunction	Multifunction	Multifunction
V	Vibration, mechanical analysis			Valve, damper, louver	
W	Weight, force		Well		
X	Unclassified	x-axis	Unclassified	Unclassified	Unclassified
Y	Event, state, or presence	y-axis		Ready, compute, convert	
Z	Position, dimension	z-axis		Driver, actuator	

Examples of instrument identification are shown in Fig. 15.5, by referring to Figs. 15.2, 15.3, and Table 15.2, the instrument identification can be determined as follows:

a. The first letter T indicates that the instrument is in temperature loop number 178. The second letter Y denotes conversion, which from the line description gives the conversion from a current of 4 to 20 mA to a pressure of 3 to 15 psi. The instrument is a discrete instrument located in the field.

b. The designation of F indicates flow, R is for recorder, and C is a controller indicating a recording flow controller in loop 97. This is an accessible computer function.

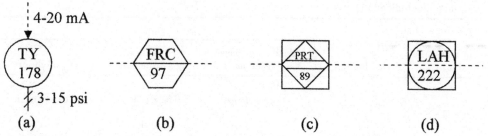

Figure 15.5 Examples of the letter and numbering codes.

 c. The letter P denotes pressure, R is recorder, and the third letter T is transmitter, giving a recording pressure transmitter in loop 89 which is located in a secondary accessible location and is a PLC function.

 d. The first letter L stands for level, A indicates alarm, and H is high, which is an alarm for high liquid levels located in loop 222 and is not accessible.

15.4 Functional Symbols

A number of functional symbols or pictorial drawings are available for most P and ID elements. A few examples are given here to acquaint the student with these elements. They have been divided into actuators, primary elements, regulators, and math functions for clarity.

15.4.1 Actuators

The first row of examples and the last three drawings shown in Fig. 15.6 are the basic sections used in some of the actuator diagrams. The other drawings show how these basic sections can be combined to form families of actuators. For instance, the hand actuator and the pneumatic actuator are shown combined with the control valve symbol to give a representation of a hand operated valve and a pneumatic operated valve in the second row. Note should also be taken of the arrows to represent the state of the valve under the system "fail" conditions.

15.4.2 Primary elements

By far the largest numbers of elements used in P and ID are the primary elements; a sampling of these elements is given in Fig. 15.7. Lettering and numbers are included in the examples.

Documentation and Symbols

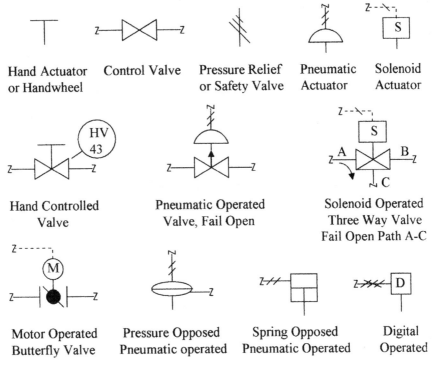

Figure 15.6 Examples of basic and actuator symbols.

15.4.3 Regulators

Typical examples of regulators and safety valves are shown in Fig. 15.8.

15.4.4 Math functions

PLCs have a large number of math functions that can be implemented using software. If these math functions are incorporated into a P and ID they will probably be executed using hardware, e.g., use of a square root to convert a pressure measurement to flow data. These functions have been symbolized; an example of the math symbols is shown in Fig. 15.9.

15.5 P and ID Drawings

All processing facilities will have a set of drawings using the standardized ISA symbols to show the plumbing, material flow, instrumentation, and control lines. The drawings normally consist of one or more main drawing depicting the facility on a functional basis with support drawings showing details of the individual functions. In a large processing plant these could run into many tens of drawings.

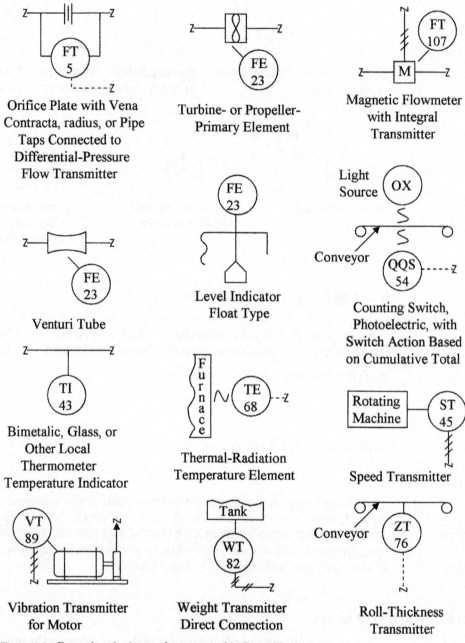

Figure 15.7 Examples of primary elements used in P and ID.

Documentation and Symbols   269

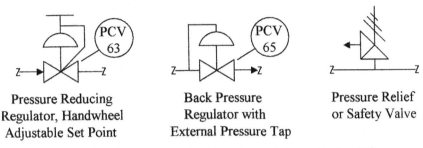

Figure 15.8 Examples of regulators and safety valve symbols used in P and ID.

Each drawing should have a part list, should be numbered, and have an area for revisions, notes, and approval signatures. It is imperative that these drawings are kept up-to-date; a few minutes taken to update a drawing can save many hours at a later date trying to figure out a problem on equipment that has been modified but whose drawings have not been updated. Figure 15.10 shows an example of a function block. The interconnection lines and instruments are clearly marked and control loops numbered. A materials list is attached with appropriate places for revisions and signatures.

Summary

This chapter introduced the documentation for alarm and trip systems, PLCs, P and IDs, and the standards developed for the symbols used in P and ID drawings.

Points of discussion in this chapter were as follows:

1. Alarm and trip systems and system documentation. The different types of sensors used in alarm system and the use of redundancy in alarm systems
2. The documentation required in PLC systems
3. The development of standards for process control symbols and drawings by ISA. The standards cover interconnection, supply lines, and the line symbols to be used

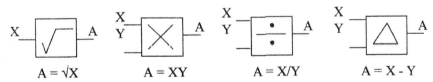

Figure 15.9 Examples of math symbols used in P and ID.

270 Chapter Fifteen

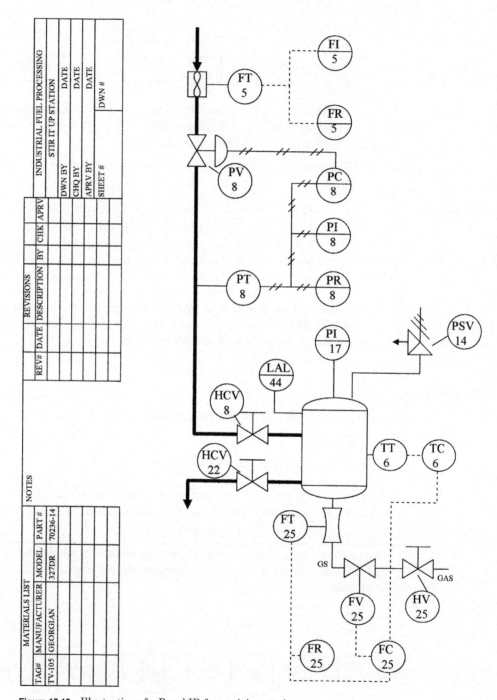

Figure 15.10 Illustration of a P and ID for a mixing station.

Documentation and Symbols

4. Symbols used for instruments, their identification, and the functional letters used with instruments and their meaning
5. Basic primary element symbols are shown and how they can be used make more complex elements
6. Examples of P and ID facility drawings and the information that should be contained in the drawings

Problems

15.1 What does the drawing in Fig. 15.11a represent?

15.2 Draw a steam supply line and attach the line indicator.

15.3 What do you understand by the symbol shown in Fig. 15.11b?

15.4 Draw a speed recorder symbol as a computer function in the field location.

15.5 Describe the symbol shown in Fig. 15.11c.

15.6 Draw an electrically operated three way valve.

15.7 What does the symbol in Fig. 15.11d represent?

15.8 Draw a solenoid operated butterfly valve which is "open" in the fail mode.

15.9 What does the symbol in Fig. 15.11e represent?

15.10 What does the symbol in Fig. 15.11f represent?

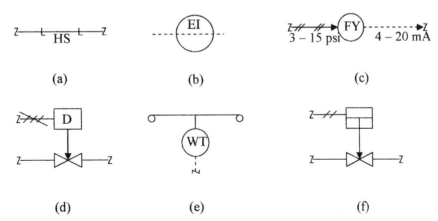

Figure 15.11 Diagrams and symbols for use with Probs. 15.1 through 15.10.

15.11 Why should documentation be kept up-to-date?

15.12 Who normally has the responsible for keeping P and ID drawings up-to-date?

15.13 Who normally has the responsibility for developing P and ID drawings?

15.14 List the information that should be contained in a P and ID drawing.

15.15 List the information that should be contained in PLC documentation.

15.16 List the information that should be contained in alarm and trip documentation.

15.17 What is the purpose of the SIS?

15.18 What are the differences between the type of sensors used in SIS and process control?

15.19 Draw the symbol of an internal pressure loaded regulator.

15.20 Draw the symbol of a pneumatic operated butterfly valve.

Appendix A

Units

A standardized system of units is required for the measurement of physical properties. Over the years, two systems of measurements have been standardized. They are the English system of units, which is still in common use in the United States, and the SI (Systéme International d'Unités) system of units. However, efforts are being made to standardize on the SI system. The SI units are sometimes referred to as the centimeter-gram-second (CGS) units and are based on the metric system but it should be noted that not all of the metric units are used. The SI system of units is maintained by the Conférence Genérale des Poids et Measures. Because both systems are in common use it is necessary to understand both system of units and to understand the relation between them. A large number of units (electrical) in use in the English system are SI units. Table A.1 gives the base units in both systems. Table A.2 gives SI units. Table A.3 gives English units. Table A.4 gives conversion between units. Table A.5 gives a list of some of the metric units that are not used in the SI system.

TABLE A.1 Base Units

Quantity	English unit	English symbol	SI unit	SI symbol
Length	Foot	ft	Meter	m
Mass	Pound (slug)	lb	Kilogram	kg
Time	Second	s	Second	s
Temperature	Rankine	R	Kelvin	K
Electric current	Ampere	A	Ampere	A
Luminous intensity	Candle	c	Lumen	lm
Angle	Degree	°	Radian	rad

TABLE A.2 SI Units Derived from Base Units

Quantity	Name	Symbol	Other unit	Base unit
Frequency	Hertz	Hz	s^{-1}	s^{-1}
Energy	Joule	J	N·m	$m^2 \cdot kg \cdot s^{-2}$
Force	Newton	N	$m \cdot kg/s^2$	$m \cdot kg \cdot s^{-2}$
Pressure	Pascal	Pa	N/m^2	$m^{-1} \cdot kg \cdot s^{-2}$
Power	Watt	W	J/s	$m^2 \cdot kg \cdot s^{-3}$
Wavelength	Meter	m	m	m
Charge	Coulomb	C	s·A	s·A
Electromotive force	Volt	V	A Ω or W/A	$m^2 \cdot kg \cdot s^{-3} \cdot A^{-1}$
Resistance	Ohm	Ω	V/A	$m^2 \cdot kg \cdot s^{-3} \cdot A^{-2}$
Conductance	Siemen	S	A/V	$m^{-2} \cdot kg^{-1} \cdot s^3 \cdot A^2$
Capacitance	Farad	F	A·s	$m^{-2} \cdot kg^{-1} \cdot s^4 \cdot A^2$
Inductance	Henry	H	Wb/A	$m^2 \cdot kg \cdot s^{-2} \cdot A^{-2}$
Magnetic flux	Weber	Wb	V·s	$m^2 \cdot kg \cdot s^{-2} \cdot A^{-1}$
Flux density	Tesla	T	Wb/m^2	$kg \cdot s^{-2} \cdot A^{-1}$
Illuminance	Lux	lx	lm/m^2	$m^{-2} \cdot cd \cdot sr$
Luminous flux	Lumen	lm	cd·sr	cd·sr
Capacity	Liter	L	dm^3	dm^3

TABLE A.3 English Units Derived from Base Units

Quantity	Name	Symbol	Unit
Energy	Foot-pound	ft·lb	$lb \cdot ft^2 \cdot s^{-2}$
Force	Pound	lb	$lb \cdot ft \cdot s^{-2}$
Pressure	Pounds per in^2	psi	$lb \cdot in^{-2}$
Power	Horsepower	hp	$lb \cdot ft^2 \cdot s^{-3}$
Specific heat	British thermal units	BTU (Btu)	$ft^2 \cdot s^{-2} \cdot °F^{-1}$
Volume	Gallon	gal	$0.1337 \, ft^3$

TABLE A.4 Conversion between English and SI Units

Quantity	English unit	SI unit
Length	1 ft	0.305 m
Mass	1 lb (slug)	14.59 kg
Weight	1 lb	0.454 kg
Volume	1 gal	3.78 L (l)
Force	1 lb	4.448 N
Angle	1 degree	$2\pi/360$ rad
Temperature	1°F	5/9°C
Energy	1 ft·lb	1.356 J
Pressure	1 psi	6.897 kPa
Power	1 hp	746 W
Heat	1 BTU	252 cal = 1055 J
Conduction	1 BTU/h ft °F	1.73 W/m K
Expansion	1 α/°F	1.8 α/°C
Specific weight	$1 \, lb/ft^3$	$0.157 \, kN/m^3$
Density	$1 \, slug/ft^3$	$0.516 \, kg/m^3$
Dynamic viscosity	$1 \, lb \, s/ft^2$	49.7 Pa·s (4.97 P)
Kinematic viscosity	$1 \, ft^2/s$	$9.29 \times 10^{-2} \, m^2/s$ (929 St)
Torque	1 lb ft	1.357 N·m

TABLE A.5 Metric Units not Normally Used in the SI System

Quantity	Name	Symbol	Equivalent
Length	angstrom	Å	1 Å = 0.1 nm
Volume	stere	st	1 st = 1 m^3
Force	dyne	dyn	1 dyn = 10 µm
Pressure	torr	torr	1 torr = 133 Pa
Energy	calorie	cal	1 cal = 4.1868 J
	erg	erg	1 erg = 0.1 µJ
Viscosity dynamic kinematic	poise	P	1 P = 0.1 Pa·s
	stoke	St	1 St = 1 cm^2/s
Conductance	mho	mho	1 mho = 1 S
Magnetic field strength	oersted	Oe	1 Oe = 80 A/m
Magnetic flux	maxwell	Mx	1 Mx = 0.01 µWb
Magnetic flux density	gauss	Gs (G)	1 Qs = 0.1 mT

Appendix B

Thermocouple Tables

The following give examples of the tables for J-, K-, S-, and T-type thermocouples. The thermocouple EMF is given in mV for 10°C degree temperature increments. The cold junction is held at 0°C. The output voltage for the different types of thermocouples may vary slightly between manufacturers.

TYPE J Iron—Constantan

	0	10	20	30	40	50	60	70	80	90
−100	−4.63	−5.03	−5.42	−5.80	−6.16	−6.50	−6.82	−7.12	−7.40	−7.66
−0	0.00	−0.50	−1.00	−1.48	−1.96	−2.43	−2.89	−3.34	−3.78	−4.21
+0	0.00	0.50	1.02	1.54	2.06	2.58	3.11	3.65	4.19	4.73
100	5.27	5.81	6.36	6.90	7.45	8.00	8.56	9.11	9.67	10.22
200	10.78	11.34	11.89	12.45	13.01	13.56	14.12	14.67	15.22	15.77
300	16.33	16.88	17.43	17.98	18.54	19.09	19.64	20.20	20.75	21.30
400	21.85	22.40	22.95	23.50	24.06	24.61	25.16	25.72	26.27	26.83
500	27.39	27.95	28.52	29.08	29.65	30.22	30.80	31.37	31.95	32.53
600	33.11	33.70	34.29	34.88	35.48	36.08	36.69	37.30	37.91	38.53
700	39.15	39.78	40.41	41.05	41.68	42.28	42.92			

TYPE K   Chromel—Alumel

	0	10	20	30	40	50	60	70	80	90
−100	−3.49	−3.78	−4.06	−4.32	−4.58	−4.81	−5.03	−5.24	−5.43	−5.60
−0	0.00	−0.39	−0.77	−1.14	−1.50	−1.86	−2.20	−2.54	−2.87	−3.19
+0	0.00	0.40	0.80	1.20	1.61	2.02	2.43	2.85	3.36	3.68
100	4.10	4.51	4.92	5.33	5.73	6.13	6.53	6.93	7.33	7.73
200	8.13	8.54	8.94	9.34	9.75	10.16	10.57	10.98	11.39	11.80
300	12.21	12.63	13.04	13.46	13.88	14.29	14.71	15.13	15.55	15.98
400	16.40	16.82	17.24	17.67	18.09	18.51	18.94	19.36	19.79	20.22
500	20.65	21.07	21.50	21.92	22.35	22.78	23.20	23.63	24.06	24.49
600	24.91	25.34	25.76	26.19	26.61	27.03	27.45	27.87	28.29	28.72
700	19.14	29.56	29.97	30.39	30.81	31.23	31.65	32.06	32.48	32.89
800	33.30	33.71	34.12	34.53	34.93	35.34	35.75	36.15	36.55	39.96
900	37.36	37.76	38.16	38.56	38.95	39.35	39.75	40.14	40.53	40.92
1000	41.31	41.70	42.09	42.48	42.87	43.25	43.63	44.02	44.40	44.78
1100	45.16	45.54	45.92	46.29	46.67	47.04	47.41	47.78	48.15	48.52
1200	48.89	49.25	49.62	49.98	50.34	50.69	51.05	51.41	51.76	52.11
1300	52.46	52.81	53.16	53.51	53.85	54.20	54.54	54.88		

TYPE S   Platinum (Rhodium 10 %)—Platinum

	0	10	20	30	40	50	60	70	80	90
0	0000	0.056	0.113	0.173	0.235	0.299	0.364	0.431	0.500	0.571
100	0.643	0.717	0.792	0.869	0.946	1.025	1.166	1.187	1.269	1.352
200	1.436	1.521	1.607	1.693	1.780	1.868	1.956	2.045	2.135	2.225
300	2.316	2.408	2.499	2.592	2.685	2.778	2.872	2.966	3.061	3.156
400	3.251	3.347	3.442	3.539	3.635	3.732	3.829	3.926	4.024	4.122
500	4.221	4.319	4.419	4.518	4.618	4.718	4.818	4.919	5.020	5.122
600	5.224	5.326	5.429	5.532	5.635	5.738	5.842	5.946	6.050	6.155
700	6.260	6.365	6.471	6.577	6.683	6.790	6.897	7.005	7.112	7.220
800	7.329	7.438	7.547	7.656	7.766	7.876	7.987	8.098	8.209	8.320
900	8.432	8.545	8.657	8.770	8.883	8.997	9.111	9.225	9.340	9.455
1000	9.570	9.686	9.802	9.918	10.035	10.152	10.269	10.387	10.505	10.623
1100	10.741	10.860	10.979	11.098	11.217	11.336	11.456	11.575	11.695	11.815
1200	11.935	12.055	12.175	12.296	12.416	12.536	12.657	12.777	12.897	13.018
1300	13.138	13.258	13.378	13.498	13.618	13.738	13.858	13.978	14.098	14.217
1400	14.337	14.457	14.576	14.696	14.815	14.935	15.054	15.173	15.292	15.411
1500	15.530	15.649	15.768	15.887	16.006	16.124	16.243	16.361	16.479	16.597
1600	16.716	16.834	16.952	17.069	17.187	17.305	17.422	17.539	17.657	17.774
1700	17.891	18.008	18.124	18.241	18.358	18.474	18.590			

TYPE T   Copper—Constantan

	0	10	20	30	40	50	60	70	80	90
−100	−3.349	−3.624	−3.887	−4.138	−4.377	−4.603	−4.817	−5.018	−5.205	−5.379
−0	0.0000	−0.380	−0.751	−1.112	−1.463	−1.804	−2.135	−2.455	−2.764	−3.062
+0	0.0000	0.389	0.787	1.194	1.610	2.035	2.467	2.908	3.357	3.813
100	4.277	4.749	5.227	5.712	6.204	6.703	7.208	7.719	8.236	8.759
200	9.288	9.823	10.363	10.909	11.457	12.015	12.575	13.140	13.710	14.285
300	14.864	15.447	16.035	16.626	17.222	17.821	18.425	19.032	19.624	20.257

Appendix C

References and Information Resources

Information Resources

There are a large number of resources for additional reading on instrumentation and process control. The internet contains a large number of Web sites that can be used as resources for more information. A list of Web site references is given below; this list is by no means complete.

Magazines

1. *Control*, www.controlmagazine.com
2. *Instrument and Control Systems*, www.icsmagazine.com
3. *Instrument and Automation News*, www.ianmag.com
4. *Sensors*, www.sensorsmag.com

Organizations

1. Institute of Electrical and Electronic Engineers, www.ieee.org
2. Instrumentation, Systems, and Automation Society, www.isa.org
3. National Institute of Standards and Technology, www.nist.gov
4. American National Standards Institute, www.ansi.org
5. National Electrical Manufactures Association, www.nema.org
6. Industrial Control and Plant Automation, www.xnet.com
7. Society of Automotive Engineers, www.sae.org/servlets/index

PLC Manufacturers

1. GE, www.geindustrial.com/cwc/gefanuc/index.html
2. Mitsubishi, www.mitsubishielectric.com/bu/automation/index
3. Rockwell, www.rockwellautomation.com
4. Siemens, www.simatic.com
5. Foxboro, www.foxboro.com
6. Honeywell, www.honeywell.com

Component suppliers

1. Texas Instrument, www.ti.com
2. National Semiconductor, www.national.com
3. DesignInfo, www.designinfo.com
4. Valves, k Controls, www.k-controls.co.uk
5. Omega Engineering, www.omega.com
6. Burr-Brown, www.burr-brown.com
7. Linear Technologies, www.linear.com/prodinfo/dnlist.html
8. Alpha, www.alphasensors.com
9. Micro Strain, www.microstrain.com
10. Entran, www.entran.com
11. Kavlico, www.kavlico.com
12. Flow Meters, www.desighinfo.com/vendors/0013.html
13. Omron, www.omron.com
14. Motorola, www.mot-sps.com
15. International Rectifier, www.irf.com
16. Siliconix, www.vishay.com/company/brands/siliconix/
17. GE, www.gesensing.com
18. Phillips, www.semiconductors.philips.com
19. Intersil Corporation, www.intersil.com
20. Heat Pipe Technology, Inc., www.heatpipe.com

Tutorial reference

1. PLC Tutor, www.plcs.net
2. Hewlett Packard, www.tmo.hp.com/tmo/iia/edcorner/English

3. Cyber Research, www.cyberresearch.com/tech/DADesign.html
4. Temperature World, www.temperatureworld.com
5. BHL, www.bhl.com
6. Macrosensors, www.macrosensors.com/primer/primer.html

References

1. Charles, A.S., *Electronics Principles and Applications*, McGraw-Hill, New York, 1999.
2. Rodger, L.T., *Digital Electronics*, McGraw-Hill, New York, 2003.
3. Ljubisa, R., *Sensor Technology and Devices*, Artech House, Norwood, MA 1994, pp. 377–456.
4. Cascetta, F. and V. Paolo, *Flowmeters: A Comprehensive Survey and Guide to Selection*, ISA, Research Triangle Park, NC, 1990.
5. Gillum, D.R., *Industrial Pressure, Level, and Density Measurement*, ISA, Research Triangle Park, NC, 1995.
6. McMillan, G.K., *pH Measurement and Control*, ISA, 1994.
7. Gary, D., *Introduction to Programmable Logic Controllers*, 2nd ed., Delmar, Albany, NY, 2002.
8. Curtis, D.J., *Process Control Instrumentation Technology*, 7th ed., Prentice-Hall, Upper Saddle River, NJ, 2003.
9. Rex, K., Jr., "Linearization of a Thermocouple," *Sensors Magazine*, Vol. 14, No. 12, 1997.
10. Davis, M., "Choosing and Using a Temperature Sensor," *Sensors Magazine*, Vol. 17, No. 1, 2000.

Appendix D

Abbreviations

Å	Angstrom
AC	Alternating current
ADC	Analog-to-digital converter
AF	Audio frequency
AM	Amplitude modulation
ANSI	American National Standards Institute
BCD	Binary coded decimal
BJT	Bipolar junction transistor
BTU	British thermal unit
C	Coulomb
CdS	Cadmium sulfide
CdSe	Cadmium selenium
CMOS	Complementary metal oxide semiconductor
CR	Control relay
DAC	Digital-to-analog converter
dB	Decibel
DIAC	Bidirectional trigger diode
DIP	Dual inline package
EMF	Electromotive force
EMI	Electromagnetic interference
F	Farad
FET	Field effect transistor
FM	Frequency modulation

FSD	Full scale deflection
GaAs	Gallium arsenide
GaAsP	Gallium arsenide phosphide
GaP	Gallium phosphide
H	Henry
HF	High frequency
HVAC	Heating ventilation and air conditioning
Hz	Hertz
IC	Integrated circuit
IEEE	Institute of Electrical and Electronics Engineers
IGBT	Insulated gate bipolar transistor
IR	Infrared
ISA	Instrument Society of America
J	Joule
K	Kelvin
LAN	Local area network
LED	Light emitting diode
LF	Low frequency
LSB	Least significant bit
LVDT	Linear velocity displacement transformer
MCT	MOS controlled transistor
MHz	Megahertz
MOS	Metal oxide semiconductor
MOSFET	Metal oxide semiconductor field effect transistor
MRE	Magnetoresistive element
MSB	Most significant bit
N	Newton
NEMA	National Electrical Manufacturers Association
NIST	National Institute of Standards and Technology
Pa	Pascal
P&ID	Pipe and identification diagram
PCM	Pulse code modulation
pF	Picofarad
PID	Proportional integral and derivative
PLA	Programmable logic array
PLC	Programmable logic controller

PPM	Pulse position modulation
PWM	Pulse width modulation
R	Rankin
RC	Resistance capacitance
RF	Radio frequency
RMS	Root mean square
RPM	Revolutions per minute
RTD	Resistance temperature device
SCR	Silicon controlled rectifier
SI	Systéme International d'Unités
SiC	Silicon carbide
SIS	Safety instrumented system
SPL	Sound pressure level
TC	Time constant
TCE	Temperature coefficient of expansion
TDM	Time division multiplex
TRIAC	Bidirectional ac switch
UPS	Uninterruptible power supply
W	Watt
WAN	Wide area network
Wb	Weber

Glossary

Absolute accuracy The accuracy stated as a definite amount, i.e., not as a percentage.

Absolute position measurement Position measured from a fixed point.

Absolute pressure Pressure measured with reference to a perfect vacuum.

Accelerometer A sensor for measuring acceleration or the rate of change of velocity.

Accuracy A measure of the difference between the indicated value and the true value.

Actuator A device that performs an action on one of the input variables of a process according to a signal received from the controller.

ADC An analog-to-digital converter that converts an analog voltage or current into a digital signal.

Alarm A warning that a variable has exceeded set limits.

Alternating current Current that flows in one direction during one half of a regular time period and the opposite direction during the other half.

Ammeter An instrument for measuring electrical current or electron flow.

Ampere The unit of current or electron flow.

Amplifier An electrical circuit that increases the magnitude of a signal.

Analog A continuously varying signal.

Aneroid barometer A barometer which uses an evacuated capsule as a sensing element.

Anticipatory action See **Derivative action**.

Aqueous solution A solution containing water.

Atmospheric pressure The pressure acting on objects on the earth's surface caused by the weight of the air in the earth's atmosphere, normally measured at sea level.

Barometer An instrument used for measuring atmospheric pressure.

Bellows A pressure sensor that converts pressure into linear displacement.

Bernoulli equation A flow equation based on the conservation of energy which includes velocity, pressure, and elevation terms.

Beta ratio The ratio of the diameter of a restriction to the diameter of the pipe containing the restriction.

Bimetallic A thermometer with a sensing element made of two dissimilar metals with different thermal coefficients of expansion.

Binary Two values, or a numbering system using the base 2.

Bit A binary digit.

Bourdon tube A pressure sensor that converts pressure to movement. The device is a coiled metallic tube that straightens when pressure is applied.

Bridge A network of passive components arranged so that small changes in one of the components can be easily measured.

British thermal unit A measure of heat energy, i.e., the amount of heat required to raise 1 lb of water 1°F at 68°F and atmospheric pressure.

Buffer amplifier A circuit for matching the output impedance of one circuit to the input impedance of another.

Buoyancy The upward force on an object floating or immersed in a fluid caused by the difference in pressure above and below the object.

Byte Eight bits of binary information.

Calorie A measure of heat energy, i.e., the amount of heat required to raise the temperature of 1 g of water by 1°C.

Capacitance A measure of a device's ability to store electrical charge.

Capacitance probe An instrument using the capacitance between two metal plates for measuring fluid level.

Capacitor A device that can store electrical charge.

Cell A simple power source that provides emf, usually by means of a chemical reaction.

Celsius One of the commonly used temperature scales.

Coefficient of heat transfer A term used in the calculation of heat transfer by convection.

Coefficient of thermal expansion A term used to determine the amount of linear expansion due to heating or cooling.

Comparator A device which compares two signals and outputs the difference.

Concentric plate A plate with a hole located at its center (orifice plate) used to measure flow by measuring the differential pressures on either side of the plate.

Conduction The movement of heat energy in a material by the transfer of energy from one molecule to another.

Conductivity probe An instrument using two electrodes to measure fluid level.

Continuity equation A flow equation which states that, if the overall flow rate is not changing with time, the flow rate past any section of the system must be constant.

Continuous level measurement A level measurement that is continuously updated.

Controlled variable The variable measured to indicate the condition of the process output.

Controller The element in a process control loop that evaluates any error of the measured variable and initiates corrective action by changing the manipulated variable.

Convection The movement of heat by the motion of warm or hot material.

Converter   A device that changes the format of a signal but not the type of energy used as the signal carrier, i.e., voltage to current.

Correction signal   The signal to the manipulated variable.

DAC   A device that converts a digital signal into an analog voltage or current.

Dead weight tester   A device for calibrating pressure-measuring devices which uses weights to provide the forces.

Decibel (dB)   A unit used to compare amplitude or power levels.

Density   The amount of mass in a unit volume.

Derivative action   Action that is proportional to the rate at which the measured variable is changing.

Dew point   The temperature at which the water vapor in a mixture of water vapor and gas becomes saturated and condensation starts.

Dielectric constant   The factor by which the capacitance between two plates changes when a material fills the space between the plates.

Differential amplifier   An amplifier that amplifies the difference between two inputs.

Digital   Signals having two discrete levels.

Dry-bulb temperature   The temperature indicated by a thermometer whose sensing element is dry.

Dynamic pressure   That part of the total pressure in a moving fluid caused by the fluid motion.

Dynamometer   An instrument used for measuring torque or power.

Eccentric plate   An orifice plate with a hole located below its center to allow for the passage of suspended solids.

Effective value   The dc voltage or dc current that would produce the same power in a load as the ac voltage or ac current being measured.

Electromagnetic flow meter   A flow-measuring device which senses the change in a magnetic field between two electrodes as a fluid flows between them.

Electromagnetism   The relationship between magnetic fields and electric current.

Electromotive force (emf)   The force that causes electrons to move, and is measured in volts.

Error signal   The difference in value between a measured signal and a set point.

Fahrenheit   One of the commonly used temperature scales.

Farad   The unit of capacitance.

Feedback   (1) The voltage fed from the output of an amplifier to the input in order to control the characteristics of the amplifier. (2) The measured variable signal fed to the controller in a closed-loop system, so that the controller can adjust the manipulated variable to keep the measured variable within set limits.

Fiber optics   The transmission of information through optical cables using light signals.

Flow nozzle A device placed in a flow line to provide a pressure drop that can be related to flow rate.

Flow rate The amount of fluid passing a given point in a given interval of time.

Flume An open-channel flow-measuring device.

Form drag The force acting on an object due to the impact of fluid.

Foundation fieldbus Process control bus used in the United States.

Free convection Movement of heat as a result of density differences.

Free surface The surface of the liquid in an open-channel flow that is in contact with the atmosphere.

Frequency The number of cycles completed in 1 s.

Gauge pressure The measured pressure above atmospheric pressure.

Gas thermometer A temperature sensor that converts temperature to pressure in a constant volume system.

Hall-effect sensor A transducer that converts a changing magnetic field into a proportional voltage.

Head Sometimes used to indicate pressure, i.e., 1 ft of "head" for water is the pressure under a column of water 1 ft high.

Heat A form of energy related to the motion of atoms or molecules.

Heat transfer The study of heat energy movement.

Henry (H) The unit of inductance.

Hertz (Hz) A measure of frequency in cycles/second.

Hot-wire anemometry A velocity-measuring device for gas or liquid flow that senses temperature changes, due to the cooling effect of gas or liquid moving over a hot element.

Humidity A term to indicate the amount of water vapor present in the air or a gas.

Humidity ratio The mass of water vapor in a gas divided by the mass of dry gas in the mixture.

Hydrometer An instrument for measuring liquid density.

Hydrostatic paradox The fact that pressure varies with depth in a static fluid, but is the same throughout the liquid at any given depth.

Hydrostatic pressure The pressure caused by the weight of static fluid.

Hygrometer A relative humidity-measuring device.

Hygroscopic A material that absorbs water and whose conductivity changes with moisture content.

Hysteresis The nonreproducibility in an instrument caused by approaching a measurement from opposite directions, i.e., going from low up to the value, or high down to the value.

Impact pressure The sum of the static and dynamic pressure in a moving fluid.

Impedance An opposition to ac current or electron flow caused by inductance and/or capacitance.

Incremental position measurement An incremental position measurement from one point to another, absolute position is not recorded, and position is lost if the power fails.

Indirect level-measuring device A device that extrapolates the level from the measurement of another variable, i.e., liquid level from a pressure measurement.

Inductance An electrical component that opposes a change in current or electron flow.

Inductor A device that exhibits inductance.

Instrument A device used to measure a physical variable.

Integral action The action designed to correct for long-term loads.

Kelvin The absolute temperature scale associated with the Celsius scale.

Kirchoff's current law The sum of the currents flowing at a node is zero.

Kirchoff's voltage law The algebraic sum of voltages around a closed path is zero.

Ladder logic The programmable logic used in PLCs to control automated industrial processes.

Lag time The time required for a control system to return a measured variable to its set point.

Laminar flow A smooth flow in which the fluid tends to move in layers.

LED Light emitting diode

Linearity A measure of the direct proportionality between actual value of the variable being measured and the value of the output of the instrument to a straight line.

Load The process load is a term used to denote the nominal values of all variables in a process that affect the controlled variable.

Load cell A device for measuring force.

Loudness A subjective quantity used to measure relative sound strength.

LVDT A linear variable transformer that measures displacement by conversion to a linearly proportional voltage.

Magnetorestrictive element (MRE) A magnetic field sensor that converts a changing magnetic field into a proportional resistance.

Manipulated variable The variable controlled by an actuator to correct for changes in the measured variable.

Measured variable The variable measured to indicate the condition of the process output.

Meniscus The convex or concave surface of a column of liquid in a tube.

Moment The effect of a force acting at a given perpendicular distance from a point.

Natural convection The movement of heat as a result of density differences.

Newtonian fluid A fluid in which the velocity varies linearly across the flow section between parallel plates.

Node A junction of three or more conductors.

Noise The term usually used to indicate unwanted or undesirable sounds.

Nutating disk meter A flow-measuring device using a disk that rotates and wobbles in response to the flow.

Offset The nonzero output of a circuit when the input is zero.

Ohmmeter An instrument used to measure resistance.

ON/OFF control A system in which a process actuator has only two positions, i.e., on and off.

Open-channel flow The flow in an open conduit (e.g., as in a ditch).

Operational amplifier A circuit used to amplify electronic signals.

Orifice plate A plate containing a hole which when placed in a pipe causes a pressure drop which can be related to flow rate.

Over pressure The term used to describe the maximum amount of pressure a gauge can withstand without damage or loss of accuracy.

Overshoot The overcorrection of the measured variable in a control loop.

Parabolic velocity distribution Occurs in laminar flow when the velocity across the cross-section takes on the shape of a parabola.

Parallel transmission Simultaneous transmission of a number of binary bits.

Pascal Pressure reading units (SI), i.e., newtons per square meter

Pascal's law The pressure applied to an enclosed fluid is transmitted to every part of the fluid.

Percent of reading The accuracy given in terms of the percentage of the reading.

Percentage full-scale accuracy The accuracy determined by dividing the accuracy of an instrument by its full-scale output taken as a percentage.

Period A fixed amount of time during which alternating current is completing one full cycle and is the inverse of the frequency in Hertz.

pH A term used to indicate the activity of the hydrogen ions in a solution, it helps to describe the acidity or alkalinity of the solution.

Phase A term used to describe the state of matter, i.e., solid, liquid, or gas.

Phons A unit for describing the difference in loudness levels.

Photodiode A sensor used to measure light intensity by measuring the leakage across a pn junction.

PID Proportional control with derivative and integral action.

P&ID Stands for piping and instrument diagrams.

Piezoelectric effect The electrical voltage developed across certain crystalline materials when a force or pressure is applied to the material.

Pitot-static tube A device used to measure the flow rate using the difference between dynamic and static pressures.

PLC Programmable logic controller.

Pneumatic System that employs gas for control or signal transmission.

Poise The measurement unit of dynamic or absolute viscosity.

Potentiometer (Pot) An adjustable resistance device.

Precision The smallest division that can be read on an instrument.

Pressure The magnitude of a force divided by the area over which it acts, i.e., psi or Pa.

Pressure differential The difference in pressure amplitudes at two locations.

Process A sequence of operations carried out to achieve a desired end result.

Process control The automatic control of certain process variables to hold them within given limits.

Processor A digital electronic computing system that can be used as a control system.

Profibus Process control bus used in Europe

Proportional action A controller action in which the controller output is directly proportional to the measured variable error.

Psychrometric chart A chart dealing with moisture content in the atmosphere.

Pyrometer An instrument for measuring temperature by sensing the radiant energy from a hot body.

Radiation The emission of energy from a body in the form of electromagnetic waves.

Range The lowest to the highest readings that can be made by a sensing device.

Rankine The absolute temperature scale associated with the Fahrenheit scale.

Rate action See **Derivative action**.

Reactance The opposition to an ac current or electron flow caused by a capacitor or an inductor.

Relative humidity The amount of water vapor present in a given volume of a gas, expressed as a percentage of the amount that would be present in the same volume of gas under saturated conditions at the same pressure and temperature.

Reluctance The opposition in a material to carrying magnetic flux, it is the magnetic equivalence to resistance.

Repeatability A measure of the closeness between several consecutive readings of a value.

Reproducibility The ability of an instrument to produce the same reading of a variable with repeated readings.

Reset action See **Integral action**.

Resistance A measure of the opposition to electron or current flow in a material.

Resistance thermometer (RTD) A temperature sensor that provides temperature readings by measuring the resistance of a metal wire (usually platinum).

Resistivity A temperature-dependent "constant" that reflects a material's resistance to electron flow.

Resistor A component that exhibits resistance.

Resolution The minimum detectable change of a variable in a measurement.

Reynolds Number A dimensionless number indicating whether the flow is laminar or turbulent.

Rotameter A flow-measuring device in which a float moves in a vertical tapered tube.

Saturated The condition when the maximum amount of a material is dissolved in another material at the given pressure and temperature conditions, i.e., water vapor in a gas.

Sealing fluid An inert fluid used in a manometer to separate the fluid whose pressure is being measured from the manometer fluid.

Segmented plate An orifice plate with a hole located so as to allow suspended solids to pass through.

Sensitivity The ratio of the change in output to input magnitudes.

Sensor A device that can convert a physical variable into a measurable quantity.

Serial transmission A sequential transmission of digital bits.

Set point The reference value for a controlled variable in a process control loop.

Signal conditioning The conversion of a signal to a format that can be used for transmission.

Single-point level measurement Indicates when a particular level has been reached.

Sling psychrometer A device for measuring relative humidity.

Smart sensor Integration of a processor directly into the sensor assembly to give direct control of the actuator and digital communication to a central controller.

Sone A unit for measuring loudness.

Sound pressure level The difference between the maximum air pressure at a point and the average air pressure at that point.

Span The difference between the lowest and highest reading for an instrument.

Specific gravity The ratio of the specific weight of a solid or liquid material and the specific weight of water, or for a gas, the ratio of the specific weight of the gas and the specific weight of air under the same conditions.

Specific heat The amount of heat required to raise a definite amount of a substance by one degree, i.e., 1 lb 1°F or 1 g 1°C.

Specific humidity The mass of water vapor in a mixture divided by the mass of dry air or gas in the mixture.

Specific weight The weight of a unit volume of a material.

Static pressure The part of the total pressure in a moving fluid not caused by the fluid motion.

Stoke The measurement unit of kinematic viscosity.

Strain gauge A sensor that converts information about the deformation of solid objects when they are acted upon by a force into a change of resistance.

Sublimation Passing directly from solid to vapor or vapor to solid.

Telemetry The electrical transmission of information over long distances usually by radio frequencies.

Temperature The term used to describe the hotness or coldness of an object.

Thermal conductivity A measure of the ability of a material to conduct heat.

Thermal expansion The expansion of a material as a result of its being heated.

Thermal time constant The time required for a body to heat or cool by 63.2 percent of the difference between the initial temperature and the aiming temperature.

Thermistor A temperature sensing element made from a metal oxide that usually has a negative temperature coefficient.

Thermocouple A temperature sensing device that uses dissimilar metal junctions to generate a voltage proportional to the differential temperature between the metal junctions.

Thermometer An instrument used to measure temperature.

Thermopile A number of thermocouples connected in series.

Time constant (electrical) The amount of time needed for a capacitance C, to discharge or charge through a resistance R, by 62.3 percent of the difference between the initial voltage and the aiming voltage; the product of RC gives the time constant in seconds.

Torque The name given to a force moment that tends to create a twisting action.

Torr The pressure caused by the weight of a column of mercury 1 mm high.

Total flow The amount of flow past a given point over some length of time.

Total pressure The sum of the static and dynamic pressures in a moving fluid.

Transducer A device that changes energy from one form to another.

Transfer function An equation that describes the relationship between the input and output of the function.

Transmission The transferring of information from one point to another.

Transmitter A device that conditions the signal received from a transducer so that it is suitable for sending to another location with minimal loss of information.

Turbine flow meter A flow-measuring device using a turbine wheel.

Turbulent flow An agitated flow in which there are random velocity fluctuations on top of the average flow.

U-tube manometer A glass tube in the shape of the letter U that is used to measure pressure or pressure differences.

Ultrasonic probe An instrument using high-frequency sound waves to measure fluid levels.

Vacuum (pressure) The amount that the measured pressure is below atmospheric pressure.

Velocity A measure of speed, and in a flow is the average speed across the flow and the direction of movement of a liquid.

Vena contracta The narrowing down of the fluid flow stream as it passes through an obstruction.

Venturi tube A specially shaped restriction in a section of pipe that provides a pressure drop which can be related to flow rate.

Viscometer (viscosimeter) An instrument for measuring viscosity.

Viscosity The term describing the resistance to flow of a fluid.

Volt The unit of electromotive force.

Voltage An electromotive force that causes electrons or a current to flow.

Voltage drop The difference in voltage between two points.

Vortex Swirling or rotating fluid motion.

Wavelength The time for an alternating source to complete a full cycle.

Weir An open-channel flow-measuring device.

Wet-bulb temperature The temperature indicated by a thermometer whose sensing element is kept moist.

Wheatstone bridge The most common electrical bridge circuit used to measure small changes in the value of an element.

Answers to Odd-Numbered Questions

Chapter 1: Introduction and review

1.1 A controlled variable is the monitored or measured output variable from a process that must be controlled within set limits. The manipulated variable is the input variable to a process that is controlled by a signal from a controller to an actuator. By controlling the input variable, the output variable is held within its set limits.

1.3 1 lb = 0.454 kg
63 kg = (63/0.454) lb = 138.77 lb

1.5 1 psi = 6.897 kPa
38.2 kPa = (38.2/6.897) psi = 5.54 psi

1.7 1 lb = 4.448 N
385 N = (385/4.448) lb = 86.55 lb

1.9 1 ft-lb = 1.356 J
27 ft-lb = (1.356 × 27) J = 36.6 J

1.11 % FSD accuracy = ±(3 × 100/120)% = ±2.5%

1.13 % FSD accuracy = ±(2 × 100/125)% = ±1.6%
% Span accuracy = ±(2 × 100/95)% = ±2.1%

1.15 % FSD accuracy = ±(3 × 100/120) kg = ±2.5 kg
% Span accuracy = ±(3 × 100/110) kg = ±2.7 kg

1.17 (a) Absolute accuracy = ±(45 × 0.5/100) fps = ±2.25 fps
(b) Absolute accuracy = ±(100 × 0.5/100) fps = ±0.5 fps

1.19 Hysteresis = ±7% FSD (see Fig. A1.1).

Chapter 2: Basic electrical components

2.1 $\lambda = c/f = \dfrac{3 \times 10^8}{230 \times 10^6}$ m = 1.3 m

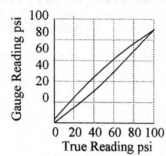

Figure A1.1 Graphs for prob. 1.19.

2.3 $R = \rho L/A$

$L = RA/\rho = \dfrac{950 \times 3.14 \times 0.16 \times 0.16}{4 \times 53}$ in $= 0.36$ in

2.5 $I = E/R = 17/133$ A $= 0.128$ A $= 128$ mA

2.7 $R = R_1 + R_2 + R_3 = 7.5 + 12.5 + 14.8$ k$\Omega = 34.8$ kΩ
$E = IR = 34.8$ k$\Omega \times 2.7$ mA $= 93.96$ V

2.9 For 7.5 kΩ $E = 20.25$ V, for 12.5 kΩ $E = 33.75$ V, for 14.8 kΩ $E = 39.96$ V
Sum of voltages = 93.96 V, i.e., same answer as prob. 2.7, therefore law holds

2.11 Volts at $A = \dfrac{16 \times 4.3}{9.7 + 4.3}$ V $= 4.91$ V

Volts at $C = \dfrac{16 \times 3.7}{8.2 + 3.7}$ V $= 4.97$ V

Difference voltage = 4.91 V − 4.97 V = −0.06 V

2.13 $X_C = 1/2\pi fC$

$f = \dfrac{1}{2\pi CX_C} = \dfrac{10^9}{2 \times 3.14 \times 3.2 \times 0.02 \times 10^6}$ Hz $= 2.84 \times 10^3$ Hz $= 2.48$ kHz

2.15 $\dfrac{1}{C} = \dfrac{1}{110\,\text{pF}} + \dfrac{1}{93\,\text{pF}} + \dfrac{1}{213\,\text{pF}}$

$C = \dfrac{110 \times 93 \times 213}{(110 \times 93) + (110 \times 213) + (93 \times 213)}$ pF $= 40.75$ pF

2.17 $L = N^2 \mu A/d$

$N = \sqrt{\dfrac{2.8 \times 5.6 \times 10^7 \times 10^4}{10^3 \times 10^2 \times 4.7 \times 3.14 \times 0.7^2}} = 1472$ turns

2.19 $L = \dfrac{L_1 \times L_2}{L_1 + L_2} = \dfrac{4.2 \times 8.7}{4.2 + 8.7}$ mH $= 2.83$ mH

Chapter 3: AC electricity

3.1 $T = CR$

$$R = \frac{T}{C} = \frac{15 \times 10^6}{0.1 \times 10^3} \, \Omega = 150 \times 10^3 \, \Omega = 150 \, \text{k}\Omega$$

3.3 $T = L/R$

$$R = \frac{L}{T} = \frac{21 \times 10^6}{12.5 \times 10^3} \, \Omega = 1680 \, \Omega = 1.68 \, \text{k}\Omega$$

3.5 $E^2 = V_R^2 + (V_L - V_C)^2$

$V_R = \sqrt{(12^2 + [16.8 - 9.5]^2)} = \sqrt{(144 - 53.29)} = \sqrt{90.71} \, \text{V} = 9.52 \, \text{V}$

3.7 $X_C = \dfrac{V_C}{I_C} = \dfrac{1}{2\pi f C}$

$$C = \frac{I_C}{V_C \times 2\pi f} = \frac{0.68 \times 10^{-3}}{9.5 \times 6.28 \times 15.74 \times 10^3} \, \text{F} = 0.72 \times 10^{-9} \, \text{F} = 0.72 \, \text{nF}$$

3.9 I at resonance $= \dfrac{E}{R} = \dfrac{12}{14 \times 10^3}$ A $= 0.86$ mA

3.11 $I_R = \dfrac{E}{R} = \dfrac{15.5}{6.5 \times 10^3} = 2.38$ mA

3.13 $I_L = \dfrac{E}{X_L} = \dfrac{15.5}{2\pi f L} = \dfrac{15.5 \times 10^3}{6.28 \times 54 \times 10^3 \times 2.3} = 19.9$ mA

3.15 $f = \dfrac{1}{2\pi\sqrt{LC}} = \dfrac{1}{6.28\sqrt{2.3 \times 6 \times 10^{-12}}}$ Hz $= \dfrac{10^6}{23.3}$ Hz $= 42.9$ kHz

3.17 I_S at resonance $= 15.5/6.5$ mA $= 2.38$ mA

3.19 $P = \dfrac{7.55^2}{2.2 \times 10^3} = 25.9$ mW

$$R = \frac{110^2 \times 10^3}{25.9} \, \text{k}\Omega = 467 \, \text{k}\Omega$$

Chapter 4: Electronics

4.1 Analog circuits are where the input and output levels are continually varying. Digital circuits are where the input and output levels are fixed, and have only two levels—high and low.

4.3 $R = \text{gain}/u = 33/5.8 \, \text{k}\Omega = 5.7 \, \text{k}\Omega$

4.5 $R = \text{gain} \times 1.3 \, \text{k}\Omega = 533 \times 1.3 \, \text{k}\Omega = 693 \, \text{k}\Omega$

4.7 The "Offset null" is used to set the output of an op-amp to zero when the differential input is zero.

4.9 Transfer ratio = output/input = 27 mV/1 μA = 27 mV/μA = 27 V/mA

4.11 $I_{out}/E_{in} = 8.5 = 100/3.5R_3$
$R_3 = 100/8.5 \times 3.5$ kΩ = 3.36 kΩ

4.13 Base number in binary is 2

4.15 0037 = 100101

4.17 111000111010 = 1110-0011-1010 = E-3-A

4.19 "1" output indicates "sourcing"

Chapter 5: Pressure

5.1 $p = \gamma h$

$$h = \frac{17.63 \times 12 \times 12}{62.4} \text{ ft} = 40.66 \text{ ft}$$

5.3 1 psf = 0.048 kPa
1038 psf = 1038 × 0.048 kPa = 49.8 kPa

5.5 Volume = 2.2 × 3.1 × 1.79 = 12.2078 ft^3

$$SW = \frac{1003 - 173}{12.2} \text{ lb/ft}^3 = 67.98 \text{ lb/ft}^3$$

SG = 67.98/64.2 = 1.09

5.7 Pressure = force/area = $\dfrac{763 \times 4}{3.14 \times 3.2 \times 3.2} = \dfrac{27 \times 4}{3.14 \times r^2}$

$r = \sqrt{\dfrac{27 \times 3.2^2}{763}}$ ft = $\sqrt{0.36}$ ft = 0.6 ft = 7.2 in

5.9 Buoyancy force = (15.5 − 8.7) × 9.8 N = 66.64 N

$$V = \frac{66.64}{9.8 \times 770} \text{ m}^3 = 0.0088 \text{ m}^3$$

$$SW = \frac{15.5 \times 9.8}{0.0088} \text{ N/m}^3 = 17.26 \text{ kN/m}^3 = 1761 \text{ kg/m}^3$$

5.11 SW = 7.38 × 62.43 lb/ft^3 = 460.7 lb/ft^3
SW = 7.38 × 1000 kg/m^3 = 7380 kg/m^3

5.13 Force = 2.9 × 1.7 × 14.3 × 12 × 12 lb = 10,151.86 lb

5.15 1.9 m = $\dfrac{1.9}{0.305}$ ft = 6.23 ft

10.3 cm = $\dfrac{10.3}{100 \times 0.305}$ ft = 0.34 ft

Pressure = (6.23 × 62.43) + (0.34 × 62.43 × 13.55) + (14.7 × 12 × 12) psfa
= 389 + 287.5 + 2116.8 psfa = 2793.3 psfa

5.17 Force = $\left(\dfrac{8.7 \times 0.305}{12}\right)^2 \times \dfrac{3.7 \times 3.14}{4}$ N = 0.14 N

5.19 Pressure = 270 × .019 psig = 5.13 + 14.7 psia = 19.83 psia

Chapter 6: Level

6.1 $p = \gamma h$

$\gamma = \dfrac{4.7 \times 144}{17}$ lb/ft^3 = 39.8 lb/ft^3

6.3 15 lb = 15 × 4.448 N = 66.72 N = 6.81 kg
V = 6.81/785 m^3 = 0.00867 m^3

6.5 Buoyancy = 17 − 3 lb = 14 lb

$V = \dfrac{14}{62.4}$ ft^3 = 0.22 ft^3

$\gamma = \dfrac{17}{0.22}$ lb/ft^3 = 77.27 lb/ft^3

6.7 Weight of liquid = 533 − 52 lb = 481 lb

$L = \dfrac{4W}{\gamma \pi d^2} = \dfrac{4 \times 481}{63 \times 3.14 \times 4.5^2}$ ft = 0.48 ft = 5.7 in

6.9 $d = \dfrac{(C_d - C_a)r}{\mu C_a}$

$\mu = \dfrac{(C_d - C_a)r}{d \times C_a} = \dfrac{(283 - 25)13 \times 12}{4 \times 31 \times 25} = 13$

6.11 Weight = $p \times A = \dfrac{32 \times 3.14 \times 3.2^2}{4}$ N = 257.4 N = 26 kg

6.13 $d = p/\gamma = \dfrac{28}{560}$ m = 0.05 m = 5 cm

6.15 $d = \sqrt{\dfrac{\delta F}{\gamma \pi h}} = \sqrt{\dfrac{3.2 \times 12}{33 \times 3.14 \times 45}}$ ft $= .0907$ ft $= 1.09$ in

6.17 $d = (C_d - C_a)\, r/\mu C_a = \dfrac{(7400 - 157) \times 2.7}{79 \times 157}$ m $= 1.58$ m

6.19 $t = d/\text{Vel} = \dfrac{2 \times 10.5 \times 0.305}{340}$ s $= 0.019$ s $= 19$ ms

Chapter 7: Flow

7.1 $Q = VA$

$V = Q/A = \dfrac{3.2 \times 4 \times 12 \times 12}{3.14 \times 7 \times 7}$ ft/s $= 11.98$ ft/s

7.3 $d = \sqrt{\dfrac{239 \times 0.1337 \times 4}{60 \times 27 \times 3.14}}$ ft $= 0.158$ ft $= 1.9$ in

7.5 $Q = \dfrac{0.73 \times 3.14 \times 23 \times 23}{4 \times 100 \times 100}$ m^3/s $= 0.03$ m^3/s $= 30$ L/s

$d = \sqrt{\dfrac{0.03 \times 4}{3.14 \times 1.66}}$ m $= 0.152$ m $= 15.2$ cm

7.7 $Q = \dfrac{3.14 \times 5.5 \times 5.5 \times 97 \times 0.1337}{4 \times 12 \times 12}$ ft^3/s $= 2.14$ ft^3/s $= 16$ gal/s

$Q(3.2) = \dfrac{2.4 \times 3.2 \times 3.2}{(3.2 \times 3.2) + (1.8 \times 1.8)}$ ft^3/s $= 1.63$ ft^3/s $= 12.19$ gal/s

$Q(1.8) = \dfrac{2.4 \times 1.8 \times 1.8}{(3.2 \times 3.2) + (1.8 \times 1.8)}$ ft^3/s $= 0.514$ ft^3/s $= 3.85$ gal/s

7.9 $m \times h = mV^2/2g$

$V = \sqrt{(2 \times 32.2 \times 273)}$ ft/s $= 132.6$ ft/s

7.11 $\dfrac{p_1}{\gamma} + \dfrac{V_1^2}{2g} + Z_1 = \dfrac{p_2}{\gamma} + \dfrac{V_2^2}{2g} + Z_2$

$0 + 0 + 17 = 0 + V_2^2/2g + 1.5$

$V_2 = \sqrt{15.5 \times 2 \times 32.2} = 31.6$ ft/s

7.13 $h_L = \dfrac{fLV^2}{2Dg}$

$h_L = \dfrac{0.027 \times 118 \times 17 \times 17 \times 12}{7 \times 2 \times 32.2}$ ft $= 24.5$ ft

7.15 $A_2 = \dfrac{3.14 \times 4.1 \times 4.1}{4 \times 12 \times 12}$ ft² = 0.092 ft²

$A_1 = \dfrac{3.14 \times 9.7 \times 9.7}{4 \times 12 \times 12}$ ft² = 0.51 ft²

$V_2 = \dfrac{28,200 \times 0.1337}{60 \times 60 \times 0.092}$ ft/s = 11.38 ft/s

$V_1 = \dfrac{11.38 \times 0.092}{0.51}$ ft/s = 2.05 ft/s

$\dfrac{p_1}{\gamma} + \dfrac{V_1^2}{2g} + Z_1 = \dfrac{p_2}{\gamma} + \dfrac{V_2^2}{2g} + Z_2$

$\dfrac{p_1}{62.4} + \dfrac{2.05 \times 2.05}{2 \times 32.2} + 0 = \dfrac{(65 + 14.7)144}{62.4} + \dfrac{11.38 \times 11.38}{2 \times 32.2} + 0$

$p_1 = (183.92 + 2.01 - 0.065)62.4 = 11,597.97$ psfa = 80.54 psia = 65.84 psig

7.17 $F = \dfrac{C_D \gamma A V^2}{2g}$

$V = \sqrt{\dfrac{4.8 \times 2 \times 32.2 \times 4 \times 12 \times 12}{0.35 \times 0.79 \times 6.3 \times 6.3 \times 3.14 \times 62.4}}$ ft/s = 9.1 ft/s

7.19 $8 \times 32/\text{rev} \times 570$ rev = 145,920 in³/h

Flow rate = 145,920/231 gal/h = 631.7 gal/h = 631.7/60 gpm = 10.53 gpm

Chapter 8: Temperature and heat

8.1 °F = (°C × 9/5) + 32 = (115 × 9/5) + 32 = 239°F

°F = (456 − 273)9/5 + 32 = 361.4°F

°F = −460 + 423 = −37°F

8.3 °C = (115 − 32) × 5/9 = 46.1°C

°C = 356 − 273 = 83°C

°C = (533 × 0.555) − 273 = 22.81°C

8.5 Heat = 3 ft³ × 62.43 lb × 15 BTU = 2809.35 BTU × 252 = 708 kcal

8.7 T increase = $\dfrac{4.3 \times 0.092 \times 50 \times 13.2 \times 17 \times 60}{1055}$ °F = 258.2 °F

8.9 Heat = $\dfrac{220 \times 3.14 \times 7 \times 7 \times 36.4 \times 12}{4 \times 12 \times 12 \times 9}$ BTU/h = 2852.11 BTU/h

304 Answers to Odd-Numbered Questions

8.11 $\dfrac{220 \times 3.14 \times 7 \times 7 \times 12(59.4 - t)}{24 \times 24 \times 9} = 30 \times 0.22(t - 23)$

$78.35(59.4 - t) = 6.6(t - 23)$

$11.87 \times 59.4 - 11.87t = t - 23$

$t = 705/12.87 = 54.8°F$

8.13 $Q = 15 \times 19 \times 0.19 \times 10^{-8}\{(125 + 460)^4 - (74 + 460)^4\} = 0.5415 \times 10^{-8}(11.712 - 8.131)10^{10}$

$Q = 194$ BTU/h

8.15 $\dfrac{2.5}{12} = 115 \times 156 \times \alpha$

$\alpha = \dfrac{2.5}{12 \times 115 \times 156} = 1.15 \times 10^{-5}/°F = 11.5 \times 10^{-6}/°F$

8.17 $R_2 = R_1(1 + \delta R\{342 \times 5/9\})$

$3074/2246 = 1 + 190\delta R$

$\delta R = 0.3686/190 = 1.94 \times 10^{-3}$ Ω/°C

8.19 $V_{out} = (1773 - 67) \times 40 \times 10^6 \times 5/9 = 0.0379$ V $= 37.9$ mV

Chapter 9: Humidity, density, and viscosity

9.1 Relative humidity = (a) 33%, (b) 20%, (c) 12%

9.3 Relative humidity = 64%

9.5 Relative humidity = 18%
Absolute humidity = 0.01 lb/lb (70 grains/lb)

9.7 25%: 0.0114 lb/lb (80 grains/lb), 95%: 0.0343 lb/lb (240 grains/lb)
Water required = 0.0343 − 0.0114 lb/lb (240 − 80 grains/lb) = 0.0229 lb/lb (160 grains/lb)

9.9 Water = 0.027 lb/lb (190 grains/lb)

9.11 Space = 14 ft³/lb × 4.7 lb = 65.8 ft³

9.13 SW = 32.2 × 1.234
$p = 32.2 \times 1.234 \times 54$ psf $= 2145.7$ psf $= 14.9$ psi

9.15 $F = \dfrac{\mu AV}{y} = \dfrac{7.3 \times 2 \times 1.2 \times 1.2 \times 14.7 \times 12}{0.11 \times 10^5}$ lb $= 0.337$ lb

9.17 pH = log 1/0.0006 = 3.22

9.19 13.2 = log 1/c

$$c = \frac{1}{10^{13.2}} = \frac{1}{1.58 \times 10^{13}} \text{ g/L} = 6.33 \times 10^{-14} \text{ g/L}$$

Chapter 10: Other sensors

10.1 $F = ma = 17 \times 21$ lb = 357 lb

10.3 Torque = $Fd = 33 \times 13$ lb-ft = 429 lb-ft

10.5 Couple = Fd
$d = 53/15$ m = 3.5 m

10.7 $w_1 \times d_1 = w_2 \times d_2$

$$d_2 = \frac{10 \times 0.5}{16} \text{ m} = 0.31 \text{ m} = 31 \text{ cm}$$

10.9 $w = 3 \times 2.7$ lb = 8.1 lb
Weight of basket = 8.1 − 6 lb = 2.1 lb

10.11 $p = f/A = \dfrac{10 \times 10^4}{75 \times 75 \times 3.14}$ Pa = 5.66 Pa

10.13 $\lambda = v/f = \dfrac{340}{13 \times 10^3}$ m = 0.026 m = 2.6 cm

10.15 Difference = $10 \log \dfrac{375}{125}$ dB = 4.77 dB

10.17 $3.83 = 10 \log d/20$
$d = 10^{0.383} \times 20 = 2.415 \times 20$ ft = 48.3 ft

10.19 Angular sensitivity = $\dfrac{360}{115 \times 16} = 0.2°$

Chapter 11: Actuators and control

11.1 Regulators are self-compensating pressure reducers. The regulators can have internal or external feedback and can use spring, weight, or external pressure for a reference.

11.3 An instrument known as a pilot-operated pressure regulator is a pressure regulator that is an externally compensated regulator; it uses an external air supply to obtain feedback amplification to enhance regulation and range.

11.5 Regulators can be spring, weight, or pressure loaded.

11.7 Electrical contactors are used to switch electrical power to high voltage/current motors and equipment and to give isolation between the low-level voltage control circuits and the high power circuits.

11.9 Optoisolators are used in low-level digital electrical circuits to give voltage isolation and ground separation between different system blocks.

11.11 A DIAC is used to set the trigger voltage level of a TRIAC or similar solid-state power switch.

11.13 The SCR is triggered on the positive half-cycle only, the TRIAC can be triggered on both the positive and negative half-cycles. The saturation voltage of the TRIAC is higher than that of the SCR.

11.15 There are five valve families in common use, they are globe, butterfly, diaphragm, ball, and rotary plug.

11.17 $C_V = Q \times \sqrt{(SG/Pd)}$

$$Pd = \left(\frac{Q}{C_V}\right)^2 \times SG = \left(\frac{1.8 \times 60}{88}\right)^2 \times \frac{78}{62.4} \text{ psi} = 1.51 \times 1.25 \text{ psi} = 1.9 \text{ psi}$$

11.19 To control the power, the maximum time constant is at half-power which is one-fourth of the cycle time.

$\frac{1}{4}$ cycle time = $\frac{1}{4} \times 1/60$ s = 4.17 ms

$5 = 12(1 - e^{-t/RC})$

From which $t = 0.54\ RC = 4.17$ ms

$C = 4.17/0.54 \times 25 \times 1000 \times 1000 = 0.31\ \mu F$

Chapter 12: Signal conditioning

12.1 The two magnetic field sensors most commonly used are the Hall effect device and the magneto resistive element.

12.3 Signals have to be conditioned to compensate for the following:
a. Sensor output signals are not always referenced to ground.
b. Sensors are temperature sensitive, i.e., the output changes with temperature.
c. Adjust the sensitivity of the sensor.
d. The variable monitored by the sensor and its output do not necessarily have a linear relationship.
e. Amplification of low-level signals for noise reduction and transmission.
f. Filtering for reduction or minimization of pickup and noise.

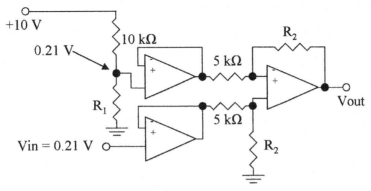

Figure A12.2 Circuit used for prob. 12.5.

12.5 To generate a 0.21 V reference voltage from 10 V (see Fig. A12.2):

$0.21(10 + R_1) = 10 \times R_1$

$R_1 = \dfrac{2.1}{9.79}$ kΩ = 0.2145 kΩ = 214.5 Ω

Feedback resistor $(R_2) = \dfrac{5 \times 10}{0.56 - 0.21}$ kΩ = 142.86 kΩ

12.7 A reference temperature for a thermocouple can be provided by a constant temperature enclosure for the cold junction, or measuring the temperature of the cold junction and correcting the output signal electrically.

12.9 The transducers used with diaphragm-type pressure sensors use strain gauges to convert the strain in the diaphragm into resistance for electrical measurement, or capacitive sensors use capacitive changes to convert the movement in the diaphragm into electrical signals.

12.11 Amplifiers with nonlinear elements in their feedback can be used for linearization, such as logarithm and antilogarithm amplifiers.

12.13 A resistor in parallel with a nonlinear element in a logarithmic amplifier will reduce the gain at the lower input voltages, and only slightly reduce the gain at the higher input voltages.

12.15 Temperature corrections can be made by using a similar element in the adjacent arm of a bridge set up as with a strain gauge. Look up tables for correction can be used, or temperature sensitive elements to control the sensitivity of the output of the sensor as in an amplifier.

12.17 Linearization of a potentiometer output controlled by a float on an angular arm, or a float with a pulley and counter balance driving a potentiometer.

12.19 The meter contacts to a RTD should be as close to the element as possible to eliminate errors due to lead resistance, and the voltage drop due to supply current flowing in the lead wires.

Chapter 13: Signal transmission

13.1 Data can be transmitted as analog signals using voltage or current levels, or as a digital transmission over hard wired connections, digital transmission can be used over fiber-optic cables or as RF signals.

13.3 RTD use two wire, three wire, or four wire connections. The two wire system is the least expensive with the four wire system the most expensive but most accurate. The three wire system uses compensation to correct for errors introduced in the wiring so that it approaches the accuracy of the four wire system at medium cost.

13.5 Two techniques are normally used to convert digital to analog signals. A resistor network can be used to convert the signals or pulse width modulation can be used.

13.7 There are several digital transmission standards, the two most common are the IEEE-488 ("1" > 2 V and "0" < 0.8 V), and the RS-232 ("1" +3 to +25 V and "0" −3 to −25 V) but in many cases these standards are being replaced by other standards.

13.9 Digital signals transfer data faster and more accurately than analog signals, are unaffected by noise, can be isolated if the ground voltage levels are different, can be transmitted over very long distances without loss of accuracy, and data can be stored.

13.11 Foundation Fieldbus has a transmission speed of 31.25 kb/s for the H1 and 100 Mb/s for the HSE.

13.13 PPM is a technique used in width amplitude modulation, to minimize the power requirements by transmitting only a pulse that is coincident with the lagging edge of the PWM signal, hence requiring less power than a PWM signal.

13.15 Amplitude modulation uses less power than frequency modulation, conserving on battery power.

13.17 Pneumatic signals are used in place of electrical signals for safety reasons, such as when there is a chance that a spark from an electrical signal could ignite combustible material, or cause an explosion in a volatile atmosphere.

13.19 There are $2^{12} - 1$ (−1 for zero when using a 12-bit DAC) or 4095 steps. The percentage resolution is 0.024%.

Chapter 14: Process control

14.1 ON/OFF action is the simplest form of control. The output variable from a process is compared to a reference, turning the control signal to the input variable to the process "on" or "off" depending on whichever is the greater.

14.3 In proportional action the amplitude of the output variable is compared to a reference, giving an output error signal with an amplitude proportional to the amount of the deviation of the variable signal from the reference signal. The error

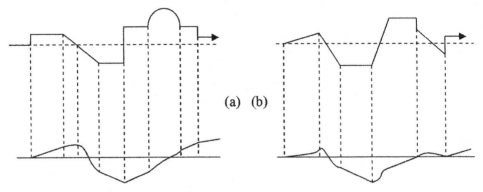

Figure A14.3 Answers to probs. 14.7 through 14.9.

signal is then used to control the input variable by an amount proportional to the amplitude of the error signal.

14.5 Derivative, rate, or anticipatory action is used to reduce the correction time that occurs with proportional action alone. Derivative action senses the rate of change of the measured variable, and applies a correction signal that is proportional to the rate of change of the measured variable only.

14.7 See Fig. A14.3(a).

14.9 See Fig. A14.3(b).

14.11 ON/OFF sensing can be used for level sensing, positioning sensing, limit sensing, HVAC, and so on.

14.13 The measured variable is the amplitude of the signal being measured. The error signal is the difference between the measured variable and the set point.

14.15 The error signal is the difference between the measured variable and the set point. The offset is that fraction of the error signal, which when amplified produces the correction signal for a change in the measured variable.

14.17 Dead-band is a set hysteresis between the turn ON level and turn OFF level in a system to prevent rapid switching between the ON and OFF points.

14.19 Derivative action is not normally used for pressure control, level control, or flow control.

Chapter 15: P and ID

15.1 A hydraulic supply line.

15.3 Discrete and inaccessible to operator voltage indicator.

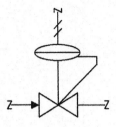

Figure A15.4 Pressure-loaded regulator.

15.5 Converter used to change a 3 to 15 psi pressure measurement to a 4 to 20 mA current measurement from a flow sensor.

15.7 Two-way digitally operated valve, which is closed in the power fail mode.

15.9 Conveyer belt with a weight measurement sensor and transmitter.

15.11 Documentation must be kept up to date to prevent time lost in maintenance, repair, and modifications, as well as to prevent catastrophic errors.

15.13 P&ID documents are normally developed by an engineering team made up of engineers from process engineering and control engineering.

15.15 Information in a PLC documentation should be

System overview and description of control process

Block diagram of units in the system

List of all inputs and outputs, destination, and number

Wiring diagram of I/O modules, address of I/O point, and rack location

Rung description, number, and function

15.17 The SIS is an alarm and trip system to alert operators or maintenance of a malfunction, to shut down a system in an orderly fashion when a malfunction occurs, or to switch failed units over to standby units.

15.19 See Fig. 15.4.

Index

Abbreviations, 283–285
Absolute accuracy, 10–11, 287
Absolute humidity, 142
Absolute position encoder, 165
Absolute position measurement, 162, 287
Absolute pressure, 69, 287
Absolute zero, 120
ac (*see* Alternating current)
Acceleration, 162
Accelerometers, 165–166
   defined, 287
   piezoelectric, 165, 166
Accuracy, 9, 10
   absolute, 10–11, 287
   defined, 9, 287
   reading, 10
Active filters, 38
Actuators, 179–198
   application considerations for, 196–198
      power devices, 197–198
      valves, 196–197
   defined, 5, 287
   electromechanical, 44
   flow control, 183–188
      ball valves, 186
      butterfly valves, 185
      globe valves, 183–185
      rotary plug valves, 186
      valve characteristics, 186–187
      valve fail safe, 187–188
      weir-type diaphragm valves, 186
   motors for, 195–196
      pneumatic feedback, 196
      servo, 195
      stepper, 195
      valve position feedback, 196
   pneumatic signals for, 205
   power control, 188–194
      electronic devices, 188–193
      magnetic control devices, 193–194
   pressure controllers, 180–183
      level regulators, 182–183
      regulators, 180–182
      safety valves, 181, 182
   symbols for, 266, 267
ADCs (*see* Analog-to-digital converters)
Air, density/specific weight of, 68
Air supply, 7
Alarm, defined, 287
Alarm and trip documentation, 261
Alarm and trip systems, 260–261
Alternating current (ac), 15, 31–45
   bridges, 39–40
   component combinations, 32–33
   current flow, 16
   defined, 287
   impedance, 32–33
   magnetic forces, 40–44
      in analog meters, 42–43
      in electromechanical devices, 43–44
      magnetic fields, 40–42
   phase change, 35–38
   R, L, and C circuits, 32–38
      phase change, 35–38
      time constants, 33–35
      voltage step, 32–33
   RC filters, 38–39
   time constants, 33–35
AM (amplitude modulation), 237
Ammeter, 287
Ampere, 287
Amplifiers, 57–58
   buffer, 55, 288
   current, 53–54
   defined, 287
   differential, 54, 289
   discrete, 48–49
   instrument, 56–57
   nonlinear, 56
   operational, 49–53, 292
Amplitude modulation (AM), 237
Analog, defined, 287
Analog circuits, 48–58, 63
   amplifier applications, 57–58
   buffer amplifiers, 55

Analog circuits (*Cont.*):
  current amplifiers, 53–54
  differential amplifiers, 54
  discrete amplifiers, 48–49
  instrument amplifiers, 56–57
  linearization in, 204
  nonlinear amplifiers, 56
  operational amplifiers, 49–53
Analog meters, 42–43
Analog signal transmission, 220–226
  current signals, 223
  noise considerations, 220–222
  resistance temperature devices, 225–226
  signal conversion, 223–224
  thermocouples, 224–225
  voltage signals, 222–223
Analog-to-digital converters (ADCs), 58, 61–63
  defined, 287
  and signal conditioning, 216
Analytical balance, 167
Anemometers, 112, 114, 115
Aneroid barometer, 287
Angular measuring devices, 215
Angular motion, 162
Angular position, 162
Aqueous solution, 287
Arc-minute, 162
Atmospheric pressure, 69, 287

Backup electrical systems, 6
Balanced cage-guided valves, 185
Ball valves, 186
Band pass filters, 38, 39
Band reject filters, 38, 39
Barometers, 79, 287
Base units of measurement, 8–9, 273, 274
Bellows, 76, 77, 207, 287
Bernoulli equation, 101, 103–105, 287
Bessel filters, 39
Beta ratio, 109, 287
Bimetallic, defined, 287
Bimetallic sensors, 128, 215
Binary, defined, 287
Binary numbers, 58–60
Bipolar amplifiers, 48, 50, 63
Bipolar junction transistors (*see* Darlington bipolar junction transistors)
Bits, 59, 288
BJTs (*see* Darlington bipolar junction transistors)
Bourdon tubes, 77–80, 207, 288
Bridges:
  alternating current, 39–40
  defined, 288
  direct current, 21–23
British thermal unit (BTU/Btu), 121, 122, 288

Bubbler devices, 93
  applications, 96
  differential, 152
Buffer amplifiers, 55, 288
Buoyancy, 71–72
  defined, 288
  as indirect liquid level measurement, 86
Butterfly valves, 185
Butterworth filters, 39
Bytes, 59, 288

Calibration:
  flow sensors, 115
  pressure sensors, 81
  temperature sensors, 137
Calorie, 121, 122, 288
Capacitance, 24–26, 288
Capacitance probe, 288
Capacitive devices:
  applications, 96
  for electrical signal conditioning, 212–213
  hygrometers, 148
  for measuring pressure, 79
  probes, 86–87, 91–92
Capacitors:
  in ac circuits, 31–38
    phase change, 35–38
    time constants, 33–35
    voltage step, 32–33
  defined, 288
  formulas, 24–25
  in parallel, 26
  in series, 26–27
Capsules, 76–77, 207
Carbon composition resistors, 18
Cell, 288
Celsius, 120, 288
Centigrade, 120
Centimeter-gram-second (CGS) units, 273
  (*See also* Systéme International D'Unités)
Chebyshev filters, 39
Chemical sensors, 170–171, 215
Clock signals, 60
Closed loop feedback systems, 245
CMOS inverter (*see* Complementary MOS inverter)
Coefficient of heat transfer, 288
Coefficient of thermal expansion, 123, 288
Comparator, 288
Complementary MOS (CMOS) inverter, 60, 61
Concentric plate, 288
Conduction, 123
  defined, 288
  heat, 125
Conductive probes, 91
Conductivity, thermal, 122–123, 295
Conductivity probe, 288

Contactors, 194
Continuity equation, 102–103, 288
Continuous control action, 243
Continuous level measurement, 288
Control circuits, PID action, 252–254
Control loop, 3–6
  block diagram of, 230
  implementation of, 249–258
    ON/OFF action electrical controller, 250–251
    ON/OFF action pneumatic controller, 249–250
    PID action control circuits, 252–254
    PID action pneumatic controller, 251–252
    PID electronic controller, 254–256
  supervisory, 231
Control parameter range, 242
Control relay, symbol for, 233
Control systems (*see* Process control)
Controlled variables, 4, 242, 288
Controllers, 5
  defined, 288
  digital, 256–257
  electronic, 254–256
  pneumatic, 249–252
  pressure, 180–183
  programmable logic, 5, 230–232
  for signal transmission, 230–235
    ladder diagrams, 232–235
    operation, 231–232
  variables measured/controlled by, 3
Convection, 123
  defined, 288
  heat, 125
  natural, 291
Converters, 5
  analog-to-digital, 58, 61–63
  defined, 202, 289
  op-amps as, 52, 53
  voltage-to-frequency, 62–63
Correction signal, 5, 289
Couple, 167
Crystals, 169
Current amplifiers, 53–54
Current flow (electricity), 16, 21
Current signals, 223

DACs (*see* Digital-to-analog converters)
Dall tube, 108, 109, 114
Darlington bipolar junction transistors (BJTs), 189, 192, 198
D'Arsonval meter movement, 43
dB (*see* Decibel)
dc (*see* Direct current)
Dead time, 242
Dead weight tester, 289
Dead-band, 242
Decibel (dB), 172, 289

Decimal number system, 58–59
Density (ρ):
  application considerations, 153
  defined, 68, 149, 289
  measuring devices for, 150–153
Derivative action, 246, 289
Dew point, 143
  defined, 289
  measuring devices for, 148
DIACs, 190, 191
Diaphragms, 75–76, 211
Dielectric constants, 24
  of common liquids, 87
  defined, 289
Differential amplifiers, 54, 289
Differential bubblers, 152
Differential control, 244
Differential pressure, 69, 70
Differential pressure gauges, 96
Differential sensing, 212–213
Digital, defined, 289
Digital circuits, 58–63
  analog-to-digital conversion, 61–63
  binary numbers, 58–60
  digital signals, 58
  linearization in, 204
  logic circuits, 60–61
Digital controllers, 256–257
Digital signal transmission, 5–6, 226–230
  foundation fieldbus and profibus, 229–230
  smart sensors, 227–228
  transmission standards, 226–227
Digital signals, 58
Digital-to-analog conversion, 235–237
  with digital-to-analog converters, 235–236
  with pulse width modulation, 236–237
Digital-to-analog converters (DACs), 235–236, 289
DIP (dual inline package), 49
Direct current (dc), 15, 16
Direct level sensing devices, 88–92
Direct reading sensors, 206–207
Discrete amplifiers, 48–49
Discrete bipolar amplifiers, 48
Displacers, 89–91, 96
Distance measuring devices, 215
Documentation, 259–261
  for alarm and trip systems, 261
  PLC, 261
  for system, 260–261
  (*See also* Symbols)
Doppler effect, 164
Drag coefficients, 107
Drift, 11
Dry particulate flow rate, 113
Dry-bulb temperature, 143, 289
Dual inline package (DIP), 49
Dual slope converters, 61

Dynamic pressure, 69, 289
Dynamometers, 169, 289

Eccentric plate, 289
Effective value, 289
Elbow, 109, 114
Electrical controllers, 250–251
Electrical signal conditioning, 207–215
   angular measuring devices, 215
   bimetallic sensors, 215
   capacitive sensors, 212–213
   distance measuring devices, 215
   float sensors, 208–211
   linear sensors, 208
   magnetic sensors, 214–215
   piezoelectric sensors, 215
   resistance sensors, 213–214
   strain gauge sensors, 211–212
   thermocouple sensors, 215
Electrical supply, 6–7
Electrical time constants, 295
Electricity, 15–16
   ac (*see* Alternating current)
   capacitance, 24–26
   inductance, 26–27
   resistance, 16–24
Electromagnetic flow meter, 111
   defined, 289
   range/accuracy of, 114
Electromagnetic interference (EMI), 6–7
Electromagnetism, defined, 289
Electromechanical devices, 43–44
Electromotive force (emf), 17, 18
   in ac circuit, 42
   defined, 289
Electronic controllers, 254–256
Electronics, 47–64
   analog circuits, 48–58, 63
      amplifier applications, 57–58
      buffer amplifiers, 55
      current amplifiers, 53–54
      differential amplifiers, 54
      discrete amplifiers, 48–49
      instrument amplifiers, 56–57
      nonlinear amplifiers, 56
      operational amplifiers, 49–53
   digital circuits, 58–63
      analog-to-digital conversion, 61–63
      binary numbers, 58–60
      digital signals, 58
      logic circuits, 60–61
   programmable logic arrays in, 64
emf (*see* Electromotive force)
EMI (*see* Electromagnetic interference)
Energy factors, 101
English system of units, 8–9, 273, 274
Error signal, 5, 242, 289

Execution mode, 232
Externally connected spring diaphragm
      regulators, 181

F (*see* Farad)
Φ (*see* Relative humidity)
Facility requirements, 6–7
Fahrenheit, 120, 289
Falling-cylinder viscometer, 155
Farad (F), 24, 289
Feedback:
   defined, 289
   pneumatic, 196
   valve position, 196
Feedback loop, 2–4, 245
FF (*see* Foundation fieldbus)
Fiber optics, 226, 289
Filtering:
   in ac circuits, 38–39
   water, 7
Fire retardant, electrical systems, 7
Fitting losses, 106
Flash converters, 61
Floats:
   applications, 95–96
   for electrical signal conditioning,
      208–211
   in sight glasses, 88–89
Flow, 99–116
   actuator control of, 183–188
      ball valves, 186
      butterfly valves, 185
      globe valves, 183–185
      rotary plug valves, 186
      valve characteristics, 186–187
      valve fail safe, 187–188
      weir-type diaphragm valves, 186
   formulas for, 102–107
      Bernoulli equation, 103–105
      continuity equation, 102–103
      flow losses, 105–107
   laminar, 100
   measurements of, 107–113
      dry particulate flow rate, 113
      flow rate, 107–111
      mass flow, 112
      open channel flow, 113, 292
      total flow, 111–112
   open-channel, 113, 292
   PID controllers for, 249
   sensors for, 114–115
      calibration of, 115
      installation of, 115
      selection of, 114–115
   terms related to, 100–102
   total, 101, 295
   turbulent, 100, 295

Flow loss formulas, 105–107
Flow meter, electromagnetic, 289
Flow nozzle, 108, 109
  defined, 290
  range/accuracy of, 114
Flow patterns, 101
Flow rate, 101, 102
  defined, 290
  measurement of, 107–111
Fluid:
  Newtonian, 291
  sealing, 294
Flumes, 113, 115, 290
FM (see Frequency modulation)
Force, 166–167
  application considerations for, 170
  defined, 166
  measuring devices for, 167–170
Form drag, 106, 107, 290
Foundation fieldbus (FF), 229–230, 290
Free convection, 290
Free surface, 290
Frequency, 15, 290
Frequency modulation (FM), 237–239
Frictional losses, 105
Functional symbols, 266–269
  actuators, 266, 267
  math functions, 267, 269
  primary elements, 266, 268
  regulators, 267, 269

γ (see Specific weight)
Gas thermometers, 130, 290
Gases:
  measuring flow of, 114, 115
  for pressure regulators, 180
  specific gravity of, 68
Gates, 60, 63
Gauge pressure, 69, 290
Gauges, pressure, 75–77
Globe valves, 183–185
Grounding, 7

H (Henry), 290
Hair hygrometers, 147
Hall device, 110, 214
Hall effect, 183
Hall-effect sensors, 164, 290
Hand-operated valves, 7
Head, 71, 290
Head loss, 106
Heat:
  conduction, 125
  convection, 125
  defined, 290

definitions related to, 121–123
radiation, 126
sensors, 170
Heat transfer:
  coefficient of, 288
  defined, 290
  formulas for, 124–126
Henry (H), 290
Hertz (Hz), 15, 290
Hexadecimal system, 59–60
High pass filters, 38, 39
High-resolution optical sensors, 174
Hot-wire anemometry, 290
Humidity, 142–149
  absolute, 142
  defined, 290
  definitions related to, 142–146
  measuring devices for, 146–149
    application considerations, 149
    dew point, 148
    hygrometers, 146–148
    moisture content, 149
    psychrometers, 148
  relative, 142, 293
  specific, 142
Humidity ratio, 142–143, 290
Hydrometers, 150–151
  defined, 290
  as direct-reading sensors, 206
  laminate, 146–147
Hydrostatic paradox, 71, 72, 290
Hydrostatic pressure, 70–71, 290
Hygrometers, 146–148
  capacitive, 148
  defined, 290
  hair, 147
  piezoelectric, 148
  resistive, 147–148
  sorption, 148
Hygroscopic, defined, 290
Hysteresis, 11, 290
Hz (see Hertz)

IEEE (Institute of Electrical and Electronic Engineers), 226
IGBTs (see Insulated gate bipolar transistors)
Impact pressure, 68, 69, 290
Impedance:
  in ac current flow, 24–25, 32–33
  in dc current flow, 25
  defined, 291
Incandescent light, 174
Inclined manometers, 73–74
Incremental optical disc, 165
Incremental position measurement, 162, 291
Indirect level-measuring device, 291
Indirect sensing devices, 92–95

Inductance, 26–27
    defined, 291
    and magnetic lines of force, 41
Induction hydrometers, 150–151
Inductors:
    in ac circuits, 31–38
        phase change, 35–38
        time constants, 33–35
        voltage step, 32–33
    defined, 291
    formulas, 26–27
    in parallel, 27
    in series, 27
Information resources, 279–281
Infra Red (IR) light-to-voltage converters, 174
Infrared absorption, 149
Infrared devices, 163–164
Installation, 7
    flow sensors, 115
    pressure sensors, 80, 81
    temperature sensors, 137
Institute of Electrical and Electronic Engineers (IEEE), 226
Instrument amplifiers, 56–57
Instrument pilot-operated pressure regulator, 182
Instrument Society of America (ISA), 7
Instrumentation, xv, 1–2
    parameters of, 9–12
    units used in, 8–10
Instruments:
    defined, 4, 291
    symbols for, 263–266
Insulated gate bipolar transistors (IGBTs), 189, 192, 198
Integral action, 247–248, 291
Intensity, light, 173
Interconnections, symbols for, 262–264
Intermediate metals, law of, 132, 133
Intermediate temperatures, law of, 132
I/O scan mode, 232
Ionization chambers, 171
Ionization gauges, 80
IR light-to-voltage converters, 174
ISA (Instrument Society of America), 7

Joules, 121

Kelvin, 120, 121, 291
Kilogram, 8
Kirchoff's current law, 21, 291
Kirchoff's voltage law, 21, 291

λ (see Wavelength)
Ladder diagrams, 232–235
Ladder logic, 291
Lag time, 242, 291
Laminar flow, 100, 291

Laminate hydrometers, 146–147
LANs (local area networks), 227
Lasers, 163–165, 174
Law of intermediate metals, 132, 133
Law of intermediate temperatures, 132
Lead compensation, 23
LED (see Light-emitting diode)
Legendre filters, 39
Level, 85–97
    formulas for, 86–87
    measurement of
        continuous, 288
        single-point, 294
    PID controllers for, 249
    sensing devices, 87–95
        direct, 88–92
        indirect, 92–95
    sensor choice, 95–97
Level regulators, 182–183
Level shifting, capacitors for, 34
Lever balance, 167
Light:
    amplitude of, 172
    application considerations for, 174, 175
    formulas for, 171–173
    intensity of, 173
    measuring devices for, 173–174
    sources of, 174
Light interference lasers, 163
Light-emitting diode (LED), 164, 165, 174, 194, 291
Light-to-frequency converters, 174
Limit switches, 232
Linear potentiometers, 20
Linear sensors, 208
Linear thermal expansion, 123, 126
Linear variable differential transformers (LVDTs), 77, 163, 208, 291
Linearity, 12, 291
Linearization, 204–205
Liquid filled thermometers, 129
Liquid in glass thermometer, 128, 129, 206
Liquids, dielectric constants of, 87
Load, 291
Load cells, 94–95, 291
Local area networks (LANs), 227
Logarithmic devices, 20
Logic circuits, 60–61
Loudness, 291
Low pass filters, 38, 39
LVDTs (see Linear variable differential transformers)

μ (see Viscosity)
Magnetic fields, 40–42
Magnetic forces (ac circuits), 40–44
    analog meters, 42–43

electromechanical devices, 43–44
   magnetic fields, 40–42
Magnetic sensors, 214–215
Magneto resistive elements (MREs), 164, 183
Magneto restrictive elements (MREs), 110, 214, 291
Maintenance, 7
Manipulated variables, 4, 291
Manometers, 73–75
   inclined, 73–74
   U-tube, 73, 295
   well, 74–75
Mass, 166
Mass flow, 103, 112
Materials, safety, 7
Math functions, symbols for, 267, 269
McLeod gauges, 80
MCT (*see* MOS-controlled thyristors)
Measured variables, 4, 242, 291
Measurement, units of (*see* Units of measurement)
Meniscus, 291
Mercury thermometers, 127–128
Metal-oxide semiconductor field effect transistors (MOSFETs), 189, 198
Metal-oxide semiconductor (MOS) devices, 188
Micro farad, 24
Microphones, 173
Microwave absorption, 149
Microwave devices, 163–164
Moisture content measuring devices, 149
Moment, 291
MOS amplifiers, 48–50, 63
MOS devices, 188
MOS-controlled thyristors (MCT), 189, 192
MOSFETs (*see* Metal-oxide semiconductor field effect transistors)
Motion:
   angular, 162
   measuring devices for, 163–166
Motors, 195–196
   pneumatic feedback, 196
   servo, 195
   stepper, 195
   symbols for, 233
   valve position feedback, 196
Moving vane, 110–111, 114
MREs (*see* Magneto resistive elements; Magneto restrictive elements)

National Institute of Standards and Technology (NIST), 9, 10, 149
Natural convection, 291
Needle valves, 185
Newton, 8

Newtonian fluid, 291
Nibbles, 59
NIST (*see* National Institute of Standards and Technology)
Node, 292
Noise, 171
   analog signal transmission, 220–222
   defined, 292
Nonlinear amplifiers, 56
NPN amplifiers, 48
Nutating disc meter, 112
   defined, 292
   range/accuracy of, 114

Offset, 11, 220, 242, 292
Offset control, 50
Ohmmeter, 292
Ohm's law, 17, 18, 25
ON/OFF control, 167, 243–244
   defined, 292
   electrical controller, 250–251
   pneumatic controller, 249–250
Op-amps (*see* Operational amplifiers)
Open channel flow, 113, 292
Open flow nozzle, 113
Operational amplifiers (op-amps), 49–53
   buffer amplifiers, 55
   defined, 292
   as instrument amplifiers, 56–57
Optical devices, 164–166
Opto-couplers, 194
Optoelectronic sensors, 175
Orifice plate, 107–109
   defined, 292
   range/accuracy of, 114
Outlet losses, 105
Over pressure, 292
Overshoot, 292

P (*see* Power transmission)
Pa (*see* Pascals)
Paddle wheels, 95, 113
Parabolic velocity distribution, 292
Parallel transmission, 292
Parshall flume, 113
Pascal's law, 72, 292
Pascals (Pa), 67, 292
Passive filters, 38–39
PD (proportional plus derivative) action, 246
Peak-to-peak (pp) values, 16
Peltier effect, 132
Percent of reading, 292
Percentage of full-scale accuracy, 292
Percentage of full-scale reading or deflection (%FSD), 10, 12, 114
Period, 292

pH, 155–157
　application considerations, 156–157
　defined, 292
　measuring devices, 156
pH meters, 155
Phase change/shift:
　in ac circuits, 35–38
　defined, 121, 292
Phons, 292
Photocells, 173
Photoconductive devices, 174
Photodiodes, 164, 165, 292
Photoemissive materials, 174
Photosensors, 175
Phototransistors, 175
Photovoltaic cells, 174
PI action (see Proportional plus integral action)
Picofarad, 24
PID control (see Proportional, integral, and derivative control)
P&ID drawings (see Pipe and identification drawings)
Piezoelectric devices:
　accelerometers, 165, 166
　electrical signal conditioning, 215
　force sensors, 169
　hygrometers, 148
　pressure gauges, 79
Piezoelectric effect, 292
Piezoresistors, 23
Pilot static tube, 109–110, 114, 292
Pilot-operated pressure regulators, 181–182
Pipe and identification (P&ID) drawings, 262–266
　defined, 292
　instrument identification, 264–266
　instrument symbols, 263–264
　interconnections, 262–263
　standardization, 262
Pirani gauge, 80
Piston flow meters, 111, 112
PLAs (programmable logic arrays), 64
PLCs (see Programmable logic controllers)
Pneumatic, defined, 293
Pneumatic actuators, 7
Pneumatic controllers:
　ON/OFF action, 249–250
　PID action, 251–252
Pneumatic feedback, 196
Pneumatic signal conditioning, 205–206
Pneumatic signal transmission, 220
Pneumatics signal conversion, 223–224
Poise, 293
Position:
　angular, 162
　application considerations, 166
　definitions related to, 161–162
　measurement of
　　absolute, 162, 287
　　devices for, 163–166
　　incremental, 162, 291
Position limit switches, 232
Positive displacement meters, 111
Potentiometers (Pot), 20, 163, 293
Pound mass, 8
Pound weight, 8
Pounds per square foot (psf), 67
Pounds per square inch (psi), 67, 68
Power control (actuators), 188–194
　electronic devices, 188–193
　magnetic control devices, 193–194
Power dissipation, 18
Power lasers, 165
Power metal-oxide semiconductor field effect transistors (power MOSFETs), 189, 192
Power switching devices, 197–198
Power transmission (P), 18–19
pp (peak-to-peak) values, 16
PPM (see Pulse position modulation)
Precision, 11, 293
Prefixes, standard (units of measurement), 9, 10
Pressure, 67–82
　absolute, 69, 287
　atmospheric, 69, 287
　defined, 67, 293
　differential, 69, 70
　dynamic, 69, 289
　formulas for, 70–73
　gauge, 69
　hydrostatic, 70–71, 290
　impact, 68, 69, 290
　as indirect liquid level measurement, 86
　measurement of, 69–70
　measuring instruments for, 73–80
　　barometers, 79
　　bellows, 77
　　Bourdon tubes, 77–79
　　capacitive devices, 79
　　capsules, 76–77
　　diaphragms, 75–76
　　gauges, 75–77
　　manometers, 73–75
　　piezoelectric pressure gauges, 79
　　vacuum instruments, 79
　PID controllers for, 249
　sensors, 80–81
　　calibration of, 81
　　installation of, 80, 81
　　selection of, 80, 81
　static, 68, 294
　total, 295
　units of, 67, 68
Pressure controllers, 180–183
　level regulators, 182–183

regulators, 180–182
  safety valves, 181, 182
Pressure differential, 293
Pressure differential sensors, 114
Pressure flow meters, 111
Pressure gauges:
  applications, 96
  piezoelectric, 79
Pressure-controlled diaphragm regulators, 180, 181
Pressure-spring thermometers, 129–130
Primary elements, symbols for, 266, 268
Probes, 91–92
Process, defined, 293
Process control, xv, 2–3, 241–257
  control loop, 3–6
  control loop implementation, 249–258
    ON/OFF action electrical controller, 250–251
    ON/OFF action pneumatic controller, 249–250
    PID action control circuits, 252–254
    PID action pneumatic controller, 251–252
    PID electronic controller, 254–256
  defined, 2, 293
  digital controllers, 256–257
  facility considerations for, 6–7
  modes of, 243–249
    derivative action, 246
    differential action, 244
    integral action, 247–248
    ON/OFF action, 243–244
    PID action, 248–249
    proportional action, 244–246
  terms related to, 242
  variables in, 1
Processor, 293
Profibus, 229–230, 293
Programmable logic arrays (PLAs), 64
Programmable logic controllers (PLCs), 5, 230–232
  defined, 293
  documentation for, 261
Proportional, integral, and derivative (PID) control, 248–249
  control circuits, 252–254
  defined, 292
  electronic controller, 254–256
  pneumatic controller, 251–252
Proportional action, 244–246, 293
Proportional plus derivative (PD) action, 246
Proportional plus integral (PI) action, 247–248
psf (pounds per square foot), 67
psi (*see* Pounds per square inch)
Psychrometers, 148
Psychrometric chart, 143–146, 293
Pulse position modulation (PPM), 237–238
Pulse width modulation (PWM), 231
  for digital-to-analog conversion, 236–237

  for telemetry, 237–238
Pyrometers, 133, 293

Quartz devices, 169

R (*see* Reynolds number)
ρ (*see* Density; Resistivity)
Radiation, 123
  defined, 293
  heat, 126
  pyrometers, 133
Radiation devices:
  applications, 93, 97
  density sensors, 152–153
Range, 10
  control parameter, 242
  defined, 293
  variable, 242
Rankine, 120, 293
RC filters, 38–39
Reactance, 293
Reading accuracy, 10
Rectilinear motion, 162
Regulators:
  pressure, 180–182
  symbols for, 267, 269
Relative humidity (Φ), 142, 293
Relays, electromechanical, 44
Reluctance, 41, 293
Repeatability, 12, 293
Reproducibility, 11, 293
Reset action, 247
Resistance, 16–24
  defined, 293
  Ohm's law, 17, 18
  power dissipation, 18
  power transmission, 18–19
  resistive sensors, 23–24
  resistivity, 17
  resistor combinations, 19–23
Resistance temperature devices (RTDs), 130
  analog signal transmission, 225–226
  defined, 293
Resistive hygrometers, 147–148
Resistive sensors, 23–24
Resistive tapes, 94
Resistive temperature detectors (RTDs), 213–214
Resistivity (ρ), 17, 294
Resistor ladder networks, 61
Resistors:
  in ac circuits, 31–38
    phase change, 35–38
    time constants, 33–35
    voltage step, 32–33
  carbon composition, 18
  defined, 294

Resistors (*Cont.*):
 in parallel, 21–23
 in series, 19–21
Resolution, 12, 294
Resonant frequency, 37, 38
Reynolds number (*R*), 100, 101, 294
Root mean square (rms), 16
Rotameters, 109, 110
 defined, 294
 as direct-reading sensors, 206, 207
 range/accuracy of, 114
Rotary plug valves, 186
Rotating disc viscometers, 155
Rotational carbon potentiometers, 20
RTDs (*see* Resistance temperature devices; Resistive temperature detectors)

Safety, 7
Safety Instrumented System (SIS), 260–261
Safety valves, 181, 182
 fail safe, 187–188
 symbols for, 267, 269
Saturated, 294
Saybolt instrument, 155
Saybolt universal viscometer, 155
SC (*see* Specific gravity)
Scan time, 232
SCRs (*see* Silicon-controlled rectifiers)
Sealing fluid, 294
Seebeck effect, 132, 133
Segmented plate, 294
Self-emptying reservoir, 182–183
Semiconductor diodes, 174
Semiconductors, 133, 137–138, 174
Sensitivity, 11, 294
Sensors, 4–5
 defined, 202, 294
 temperature sensitivity of, 205
Serial transmission, 294
Servo motors, 195
Set point, 4, 242, 294
SI (*see* Systéme International D'Unités)
Sight glasses, 88–89, 206
Signal conditioning, 201–216
 and A-D conversion, 216
 defined, 294
 direct reading sensors, 206–207
 electrical, 207–215
  angular measuring devices, 215
  bimetallic sensors, 215
  capacitive sensors, 212–213
  distance measuring devices, 215
  float sensors, 208–211
  linear sensors, 208
  magnetic sensors, 214–215
  piezoelectric sensors, 215
  resistance sensors, 213–214
  strain gauge sensors, 211–212
  thermocouple sensors, 215
 linearization, 204–205
 pneumatic, 205–206
 sensor output characteristics, 202–203
 temperature correction, 205
 visual display conditioning, 206–207
Signal conversion, 223–224
Signal integration, 34
Signal transmission, 219–239
 analog, 220–226
  current signals, 223
  noise considerations, 220–222
  resistance temperature devices, 225–226
  signal conversion, 223–224
  thermocouples, 224–225
  voltage signals, 222–223
 controllers for, 230–235
  ladder diagrams, 232–235
  operation, 231–232
 digital, 226–230
  foundation fieldbus and profibus, 229–230
  smart sensors, 227–228
  transmission standards, 226–227
 digital-to-analog conversion, 235–237
  with digital-to-analog converters, 235–236
  with pulse width modulation, 236–237
 pneumatic, 220
 telemetry, 237–239
  frequency modulation, 238–239
  width modulation, 237–238
Signals:
 analog, 61
 clock, 60
 correction, 5, 289
 digital, 58
 error, 5, 242, 289
Silicon absolute pressure gauge, 79
Silicon diaphragms, 75–76
Silicon pressure sensors, 80
Silicon-controlled rectifiers (SCRs), 188–190, 198
Sine waves, 16
Single-ended sensing, 212
Single-point level measurement, 294
Single-pole single-throw double-break contactor, 194
SIS (*see* Safety Instrumented System)
Sling psychrometer, 294
Slug, 8, 68
Smart sensors, 227–228, 294
Smoke detectors, 170–171, 215
Sone, 294
Sorption hygrometers, 148
Sound:
 application considerations for, 174, 175
 formulas for, 171–173
 measuring devices for, 173–174

Sound level meters, 173
Sound pressure levels (SPLs), 172, 294
Sound waves, 172
Span, 10, 220, 294
Specific gravity (SC):
 of common materials, 68
 defined, 68, 149, 150, 294
Specific heat, 122, 294
Specific humidity, 142, 294
Specific weight (γ), 68, 149, 294
Speed, 162
Split body valves, 185
SPLs (see Sound pressure levels)
Spring transducers, 168
Spring-controlled regulators, 180
Sprinklers, 170
Standard prefixes (units of measurement), 9, 10
Standards, 7
 and accuracy determination, 9, 10
 digital transmission, 226–227
 for symbols, 262
Static pressure, 68, 294
Stepper motors, 195
Stoke, 295
Strain gauges, 22–24
 defined, 295
 electrical signal conditioning, 211–212
 for force measurement, 169
 range/accuracy of, 114
Sublimation, 121, 295
Successive approximation, 61
Symbols, 262–270
 functional, 266–269
  actuators, 266, 267
  math functions, 267, 269
  primary elements, 266, 268
  regulators, 267, 269
 for ladder diagrams, 232–234
 for pipe and identification (P and ID) diagrams, 262–270
  for instrument identification, 264–266
  instrument symbols, 263, 264
  interconnections, 262–264
  standardization, 262
Systéme International D'Unités (SI), 8–9, 273, 274

Taguchi sensors, 171, 215
Telemetry, 237–239
 defined, 237, 295
 frequency modulation, 238–239
 pulse width modulation, 237–238
Temperature, 119, 295
 compensating for sensitivity to, 205, 207
 definitions related to, 120–121
 dry-bulb, 143, 289
 formulas for, 124
 measuring devices for, 127–134

  pressure-spring thermometers, 129–130
  resistance temperature devices, 130
  semiconductors, 133
  thermistors, 131
  thermocouples, 131–134
  thermometers, 127–128
 PID controllers for, 249
 and resistivity, 17
 sensors for, 134–138
  calibration of, 137
  installation of, 137
  protection of, 137–138
  range and accuracy of, 134, 135
  selection of, 134
  thermal time constant for, 134–137
 wet-bulb, 143, 296
Temperature scales, 120, 124
Texas Instruments, 174
Thermal conductivity, 122–123, 295
Thermal energy, 122
Thermal expansion, 295
 coefficient of, 123, 288
 definitions related to, 123
 formulas for, 126–127
 linear, 123, 126
 volume, 123, 126–127
Thermal time constant, 134–137, 295
Thermistors, 131, 295
Thermocouples, 131–134
 analog signal transmission, 224–225
 defined, 295
 electrical signal conditioning, 215
 tables for, 277–278
Thermohydrometers, 150
Thermometers, 127–128
 bimetallic strips, 128
 defined, 295
 gas, 130, 290
 liquid in glass, 128
 mercury, 127–128
 pressure-spring, 129–130
 resistance, 293
 resistance temperature devices, 130
 use of term, 127
 vapor-pressure, 129–130
Thermopiles, 132, 133, 295
Thompson effect, 132
Three-position globe valves, 184–185
Time constants:
 ac circuits, 33–35
 electrical, 295
 thermal, 134–137, 295
Time division multiplexing, 229
Torque, 167
 application considerations for, 170
 defined, 295
 measuring devices for, 167–170

Torque wrenches, 169–170
Torr, 295
Total flow, 101
   defined, 295
   measurement of, 111–112
Total pressure, 295
Total vacuum, 69
Transconductance, 49
Transducers, 5
   defined, 202, 295
   microphones, 173
Transfer function, 295
Transformers, 42
Transient, 242
Transmission:
   defined, 295
   parallel, 292
Transmission standards, 226–227
Transmitters, 5–6, 220, 295
TRIACs, 188, 190–192, 198
Turbine flow meter, 110
   defined, 295
   range/accuracy of, 114
Turbulent flow, 100, 295
Two-way globe valves, 184–185

Ultrasonic devices:
   for distance measurement, 163–164
   probes, 92, 97, 295
Uninterruptible power supplies (UPSs), 6
Units of measurement, 8–10, 273–275 (*See also specific topics, e.g.:* Pressure)
UPSs (uninterruptible power supplies), 6
U-tube manometer, 73
   defined, 295
   as direct-reading sensors, 206

Vacuum instruments, 79
Vacuum (pressure), 69, 73, 296
Valves, 196–197
   characteristics of, 186–187
   fail safe for, 187–188
   hand-operated, 7
   position feedback for, 196
   safety, 181, 182, 267, 269
   selecting, 197
   sizing of, 186–187
   symbols for, 267, 269
   (*See also specific types*)
Vapor-pressure thermometer systems, 129–130
Variable range, 242
Variables:
   controlled, 4, 242, 288
   manipulated, 4, 291
   measured, 4, 242, 291
   in process control, 1
Velocity, 100, 162, 296
Velocity meters, 112
Vena contracta, 296
Venturi tube, 108, 114
   defined, 296
   range/accuracy of, 114
Vibration, 162
Vibration sensors, 151, 162, 166
Viscometers (viscosimeters), 154–155, 296
Viscosity (μ), 100, 153–155
   defined, 153, 296
   measuring instruments for, 154, 155
Visual display conditioning, 206–207
Volt, defined, 296
Voltage:
   defined, 296
   Kirchoff's first law, 21
Voltage divider potentiometers, 20
Voltage dividers, 19
Voltage drop, 296
Voltage signals, 222–223
Voltage step, 32–33
Voltage vectors, 36
Voltage-to-current converters, 53
Voltage-to-frequency converters, 62–63
Voltmeters, 23
Volume flow rate, 103
Volume thermal expansion, 123, 126–127
Vortex, 296
Vortex flow meters, 111, 114

WANs (wide area networks), 227
Water, pH of, 155
Water supply, 7
Wavelength (λ), 16, 296
Weight, 167, 169
Weight-controlled regulators, 180
Weir, 113, 296
Weir diaphragm valves, 186
Well manometers, 74–75
Wet-bulb temperature, 143, 296
Wheatstone bridge, 21–22, 296
Wide area networks (WANs), 227
Width modulation, 237–238
Wire-wound slider type potentiometers, 20
Words (electronic), 59

X-rays, 173

CPSIA information can be obtained
at www.ICGtesting.com
Printed in the USA
BVOW09*2010060617
486201BV00011B/87/P